T0275661

Silicon Photonics

Silicon Photonics
Fueling the Next Information Revolution

Daryl Inniss

Roy Rubenstein

AMSTERDAM • BOSTON • HEIDELBERG • LONDON
NEW YORK • OXFORD • PARIS • SAN DIEGO
SAN FRANCISCO • SINGAPORE • SYDNEY • TOKYO

Morgan Kaufmann is an imprint of Elsevier

Morgan Kaufmann is an imprint of Elsevier
50 Hampshire Street, 5th Floor, Cambridge, MA 02139, United States

Copyright © 2017 Daryl Inniss and Roy Rubenstein. Published by Elsevier Inc. All rights reserved.

No part of this publication may be reproduced or transmitted in any form or by any means,
electronic or mechanical, including photocopying, recording, or any information storage and
retrieval system, without permission in writing from the publisher. Details on how to seek
permission, further information about the Publisher's permissions policies and our arrangements
with organizations such as the Copyright Clearance Center and the Copyright Licensing Agency,
can be found at our website: www.elsevier.com/permissions.

This book and the individual contributions contained in it are protected under copyright by the
Publisher (other than as may be noted herein).

Notices
Knowledge and best practice in this field are constantly changing. As new research and
experience broaden our understanding, changes in research methods, professional practices, or
medical treatment may become necessary.

Practitioners and researchers must always rely on their own experience and knowledge in
evaluating and using any information, methods, compounds, or experiments described herein. In
using such information or methods they should be mindful of their own safety and the safety of
others, including parties for whom they have a professional responsibility.

To the fullest extent of the law, neither the Publisher nor the authors, contributors, or editors,
assume any liability for any injury and/or damage to persons or property as a matter of products
liability, negligence or otherwise, or from any use or operation of any methods, products,
instructions, or ideas contained in the material herein.

British Library Cataloguing-in-Publication Data
A catalogue record for this book is available from the British Library

Library of Congress Cataloging-in-Publication Data
A catalog record for this book is available from the Library of Congress

ISBN: 978-0-12-802975-6

For Information on all Morgan Kaufmann publications
visit our website at https://www.elsevier.com

Publisher: Todd Green
Acquisition Editor: Todd Green
Editorial Project Manager: Charlie Kent
Production Project Manager: Priya Kumaraguruparan
Cover Designer: Vicky Pearson Esser

Typeset by MPS Limited, Chennai, India

This body of work is dedicated to my wife and our wonderful children.

Daryl Inniss

To my mother and my late father. He would have loved to discuss the issues raised in this book.

Roy Rubenstein

CONTENTS

We started writing this book in late 2014. At the time, silicon photonics was in a quiet period. The excitement that greeted the early announcements and later a spate of silicon photonics acquisitions had, by then, been replaced by pragmatism as the industry understood not only the technology's merits but also its challenges. Many in the industry were keeping a watching brief while others remained skeptical, questioning whether silicon photonics was a viable technology [1,2].

Silicon photonics' quiet period has ended. As we complete the book in late 2016, the year has seen more players enter the marketplace with silicon photonics' products, more acquisitions, and even a successful initial public offering, by Acacia Communications.

However, in one sense the larger picture has not changed: people have long recognized the potential of silicon, and there have also been voices of pragmatism. Take this view from Simon Sherrington, talking in 2009 about his *Light Reading* report on silicon photonics: "Silicon Photonics is changing the way vendors do photonic integration and has the potential to disrupt the supply chain." [3] And a *Nature Photonics* paper did an important job to dispel some of the myths that were being associated with silicon photonics. The paper also stated this: "The bottom line is that individual silicon devices will probably not outperform single devices based on other material platforms, except in a few particular areas. The applications that will benefit most from the silicon platform are those that require many devices to be strung together into a complex system, just like in electronics." [4] In late 2016 both statements remain true.

Silicon photonics is coming to market at a time of momentous change. One significant trend is the rise of the Internet content providers and the developments taking place in the data center. These developments are having a knock-on effect for the communications service providers—telcos—which are now undertaking their own transformation. A second change is the end of Moore's law. The chip industry is currently grappling with a wave of company consolidations, while the end of Moore's law will have far-reaching consequences. Meanwhile, the optical industry faces its own issues as the bandwidth-carrying capacity of fiber starts to be approached. Optical fiber has been seen as a communications medium of near-boundless capacity. The growth in Internet traffic, the use of smartphones, and subscribers' appetite for video means that is no longer true.

Each of these developments—the data center, the end of Moore's law, and the looming capacity crunch—is significant in its own right. But collectively they signify a need for new thinking for chips, optics, and systems, as well as new business opportunities and industry change. Silicon photonics is arriving at a propitious time.

Despite this, the optical industry still has questions regarding the significance of silicon photonics. Meanwhile, for the chip industry optics remains a science peripheral to their daily concerns. This too will change.

As implied by its name, silicon photonics is set to influence both industries. For the optical industry, the technology will allow designs to be tackled in new ways. For the chip industry, silicon photonics may be a peripheral if interesting technology, but it will impact chip design. Silicon photonics may have hurdles to overcome, but it is a technology that no one will be able to ignore.

We felt the timing was right for a book that synthesizes the significant changes taking place in the datacom, telecom, and semiconductor industries and explains the market opportunities that will result and the role silicon photonics can play.

We have cast a wide net across these industries for a reason: we see it as a vital exercise to understand the significance of silicon photonics. Indeed, the book shows that silicon photonics will be a key technology for a post-Moore's law era, and we argue that it will be the chip industry, not the photonics industry, that will drive optics.

THE REASONING FOR THE BOOK AND ITS ORGANIZATION

Let us start by saying what this book is not. It is not a traditional textbook—there is almost no math. Nor is it a compendium of the latest research work of the leading academics in the field. It is also not a how-to design book. Wonderful examples of such books exist. Nor is it a detailed market research report.

Instead, we have set ourselves a wider brief to look across important industries to tell the story of a key technology that is coming to market. To tell this story, we have broadened the context not just for the optical community but for the chip industry. The book is deliberately written with the assumption that not all the readership is familiar with optics or with the chip industry. We have brought in the voices of key silicon photonics luminaries who have played an important role in bringing the technology to market to tell some of their stories. These individuals have thought deeply about the technology and its likely ramifications.

The book focuses on the telecom and datacom industries, which are and will remain the primary markets for silicon photonics for the next decade at least. But we also note other developments where silicon photonics can play an important role.

A work of this nature must include some technical topics in optical communications, networking, and systems architectures. We have aimed to write these at a level that details what is needed without taking the reader on a detour. Interested readers will find appendices which give a more detailed discussion.

The main audiences for this book include design engineers of components—chips and optical components— as well as systems designers for a range of devices, from optical modules to telecom and datacom equipment. It is also written for sales and marketing executives who want to understand the broader developments in their industry, the changes taking place, and the key technologies. We have also targeted the book at press relations and media executives working in these industries.

Lastly, one of the silicon photonics luminaries, Lionel Kimerling, professor of materials science and engineering at MIT, told us how he is spending most of his time working with AIM Photonics, a US public–private venture established to advance the manufacturing of silicon photonics. Professor Kimerling is putting together educational material to help attract individuals to pursue a career in silicon photonics. Much of the technology is in place, he says; what is required is to make it accessible to people. "I don't have 40 more years in the industry, but I could influence the next 40 years by creating these instructional

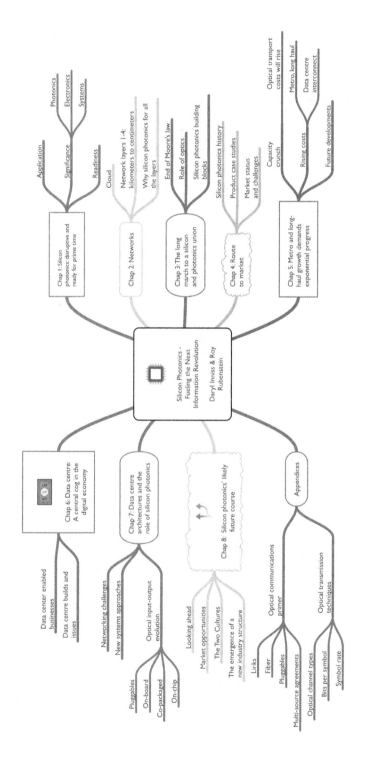

Silicon Photonics - Fueling the Next Information Revolution

Daryl Inniss & Roy Rubenstein

Chap 1: Silicon photonics: disruptive and ready for prime time
- Application
- Significance
  - Photonics
  - Electronics
  - Systems
- Readiness

Chap 2: Networks
- Cloud
- Network layers 1-4: kilometers to centimeters
- Why silicon photonics for all the layers

Chap 3: The long march to a silicon and photonics union
- End of Moore's law
- Role of optics
- Silicon photonics building blocks
- Silicon photonics history

Chap 4: Route to market
- Product case studies
- Market status and challenges

Chap 5: Metro and long-haul growth demands exponential progress
- Capacity crunch
- Rising costs
  - Optical transport costs will rise
  - Metro, long haul
  - Data centre interconnect
- Future developments

Chap 6: Data centre: A central cog in the digital economy
- Data center enabled businesses
- Data centre builds and issues

Chap 7: Data centre architectures and the role of silicon photonics
- Networking challenges
- New systems approaches
- Optical input-output evolution
  - Pluggables
  - On-board
  - Co-packaged
  - On-chip

Chap 8: Silicon photonics' likely future course
- Looking ahead
- Market opportunities
- The Two Cultures
- The emergence of a new industry structure

Appendices
- Optical communications primer
  - Links
  - Fiber
  - Pluggables
  - Multi-source agreements
- Optical transmission techniques
  - Optical channel types
  - Bits per symbol
  - Symbol rate

materials and career paths, and getting roadmap consensus that can drive the industry," says Kimerling.

In a very modest way, we hope this book also plays a role in realizing Professor Kimerling's vision and is read by students considering their engineering options.

The chart shown on the previous page summarizes the book's content at a glance.

Please visit the book companion website http://booksite.elsevier.com/9780128029756/ for full color versions of the figures in this book.

REFERENCES

[1] Silicon photonics a viable technology? Datacenter J. <http://www.datacenterjournal.com/silicon-photonics-viable-technology/>; July 6, 2016.

[2] Knocking of silicon photonics now mainstream. Fibereality, <http://fibereality.com/blog/knocking-of-silicon-photonics-now-mainstream/>; April 2015.

[3] Silicon carves out its niche as an optical material. Fibre Systems Europe; March–April 2009, pp. 18–21.

[4] Baehr-Jones T, et al. Myths and rumours of silicon photonics. Nat Photonics April 2012;6:206–8.

ACKNOWLEDGMENTS

We are indebted to the many people who have contributed to the creation of this book.

First of all, we would like to express deep thanks to Lance Leventhal without whom this work would not exist.

We would also like to express our gratitude to the silicon photonics luminaries who agreed to be interviewed: Andrew Rickman, Mario Paniccia, Graham Reed, Richard Soref, Chris Doerr, Philippe Absil, Joris Van Campenhout, John Bowers, Mehdi Asghari, Keren Bergman, Lionel Kimerling, Sanjay Patel, Young-Kai Chen, and Peter De Dobbelaere. We would also like to thank the many telecom and datacom experts we interviewed, all of whom contributed generously with their knowledge and their time.

We are grateful to the industry analysts who allowed us to cite their work and shared their insights, in particular Vladimir Kozlov, Dale Murray, and John Lively at LightCounting and Julie Kunstler, Ian Redpath, and Matt Walker at Ovum and Ron Kline and Mark Newman who were formerly at Ovum. We would also like to point out our involvement with these two market research companies. Inniss was until early 2016 vice president and practice leader of Ovum's components practice, while Rubenstein is a consultant at LightCounting.

We would like to express our gratitude to Jean Atelsek for her tireless work and invaluable comments while editing the book. Thanks also to Martin Rubenstein for his careful readings and suggestions improved the book. In turn, we would like to thank the following for their comments and feedback while the book was being written: Brandon Collings, Joris Van Campenhout, Gareth Spence, Victoria McDonald, Martin Hull, Eric Hall, Glenn Wellbrock, Julie Kunstler, Karen Liu, Vladimir Kozlov, Ioannis Tomkos, Siddharth Sheth, Peter Winzer and James Kisner.

In turn, we would like to thank the publisher Todd Green at Elsevier for commissioning our book and for his help, and Charlotte

Kent for her guidance and for keeping us on course. We are delighted that Elsevier's Morgan Kaufmann is the publisher of our book.

We also thank Alice White who gave sage advice at the start of this project and Gudmund Knudsen, who graciously supported completing the book.

We would also like to add our own personal thanks.

I [Inniss] would like to thank Ovum, where I was employed when this project started. I am in debt to the wonderful analysts there who helped form the foundation for much of the story told here. I would also like to thank OFS, where I have been employed these last 12 months. While the opinions expressed here are mine alone and not those of my employer, my work at OFS has further helped inform me about the silicon photonics market, optical communication, and where silicon photonics can be a valuable technology. I am in debt to the organization for its generosity in allowing me to complete this manuscript.

As editor of Gazettabyte, I [Rubenstein] would also like to express my gratitude to the sponsors of Gazettabyte: ADVA Optical Networking, Ciena, Finisar, Infinera, Intel, LightCounting, Nokia, and Oclaro. Without such sponsors, many of whom have backed Gazettabyte since its start, the online publication would not exist. Through Gazettabyte, it has been possible to record the ongoing developments in datacom and telecom that have been a valuable resource in charting the progress of silicon photonics. Lastly, I have over the last 23 years interviewed individuals in the chip and telecoms industries. I am continually reminded of people's generosity in taking time from their busy schedules to discuss technologies and markets. It is a wonderful privilege to have a front row seat on leading-edge products and technologies and to be briefed by experts. I would like to take this opportunity to thank them all.

Daryl Inniss and Roy Rubenstein

CHAPTER *1*

Silicon Photonics: Disruptive and Ready for Prime Time

This is the interesting thing about technology, you never really know how successful it will be.

Vladimir Kozlov [1]

There is a difference between a viable technology and the commercial application of it.

Mario Paniccia [2]

1.1 INTRODUCTION

In the 1950s the world welcomed the rise of the electronic transistor, which ultimately led to the popularization of the computer. In the 1990s optical technology enabled the exponential growth of data transmission, connecting computers globally, which gave rise to the democratization of the Internet and the World Wide Web. The next step in the journey of the digital economy—the application of optics to electronic processes and vice versa—is silicon photonics.

Optics—the use of light to send signals through a transparent path—is playing an increasingly important role in communications across a vast scale of distances. For two decades or more, it has allowed the networks of the telecommunications operators—the telcos—to cope with huge annual growth in data traffic. Now photonics is also playing a central role in the data center, where Internet data is received, processed, and distributed.

Silicon photonics can be viewed in several ways. From an optical component industry perspective, it is the most recent technology to join several established technologies used to make optical devices. This is a valid but narrow viewpoint, because silicon photonics is much more than that.

Silicon Photonics. DOI: http://dx.doi.org/10.1016/B978-0-12-802975-6.00001-6
Copyright © 2017 Daryl Inniss and Roy Rubenstein. Published by Elsevier Inc. All rights reserved.

Silicon photonics enables optical devices to be made on a silicon substrate and fabricated in a chip facility. The resulting devices are starting to be adopted by the optical industry, but the technology's commonalities with the much larger semiconductor industry is raising its profile among the chip giants. It is a technology the chip industry recognizes it will need, to tackle input–output bottlenecks in its more complex chips. The adoption of silicon photonics by the semiconductor industry will have far-reaching consequences, as is explained in the book.

This chapter introduces silicon photonics and addresses its significance. The status of silicon photonics and whether it has reached its tipping point are also discussed. Two other points are tackled briefly: its market opportunities, and whether silicon photonics is disruptive. This chapter highlights key issues and themes that are expanded upon in the book.

1.2 SILICON PHOTONICS: AN INTRODUCTION

Silicon photonics luminary Professor John Bowers of the University of California, Santa Barbara describes silicon photonics as bringing CMOS processing to optics.

CMOS—short for complementary metal-oxide semiconductor—has been the bedrock technology of integrated circuits for decades, and chip-making is one of the most advanced mass volume manufacturing processes ever developed.

Hundreds or thousands of chips, millimeters in size, are processed in parallel on a single silicon wafer measuring 300 mm (12 in.) in diameter. The bigger the wafer, the more devices can be made on it and the better the economics of chip-making. Silicon wafers are processed in chip fabrication plants, known as fabs, that run 24 hours a day, 365 days a year. Modern chip fabrication plants are hugely expensive factories, costing billions of dollars.

CMOS transistors made on wafers in these plants now have feature sizes as small as 14 nm, a fraction of the width of a human hair. Feature size refers to a key dimension of a transistor. The continual reduction in feature size by the chip industry has enabled ever more transistors to be crammed onto a chip, the consequences of which are described by Moore's law. Based on an observation by Gordon

E. Moore, the law states that the complexity of integrated circuits doubles every 18 months (later amended to every 24 months).

Silicon photonics aims to piggyback on the huge semiconductor industry—its know-how and the vast investments it has made over decades. Silicon photonics is not the same as the CMOS process used to make chips. Silicon photonics involves creating, processing, and detecting light and, not surprisingly, has its own manufacturing requirements that differ from those used to make electronic chips. These processes include not just the wafer processing to make the photonic circuits but also custom circuit testing equipment and device packaging.

But the benefits the semiconductor industry can bring to photonics are unquestionable. The chip-making process can be used to make efficient light pipes—waveguides—that direct the light between optical functions. The precision manufacturing of chip-making improves photonic device optical performance and device yields, and large silicon wafers benefit the economics of component making. The chip industry also brings packaging and testing benefits, as well as a sophisticated design tool environment.

Silicon has a key shortfall, however: it does not lase because it does not give off light—photons—when driven with electrons, a consequence of its electronic structure. Here, optical materials such as indium phosphide and gallium arsenide—known as III-V compounds based on the columns in the periodic table—are needed to provide a silicon photonic circuit's light source, a topic discussed in Chapter 3, The Long March to a Silicon-Photonics Union.

The same applies to detecting light—converting photons to electrical current using a photodetector circuit. Silicon needs help, and here the element germanium is used. The chipmakers have already been down this path of adding materials to advance CMOS, so it is not a deal-breaker for silicon photonics. But as will be explained in Chapter 3, The Long March to a Silicon-Photonics Union, it is not trivial either, and several approaches are being pursued by the silicon photonics players.

Another distinction of silicon photonics is that, unlike chip-making, it uses much larger feature sizes. The minimum size of silicon photonics waveguides—the optical equivalent of wires—is governed by the light's transmission wavelength. The light is in the infrared part of the electromagnetic spectrum; its wavelength is several orders of magnitude larger than an electron, which means that much larger feature sizes and hence older CMOS processing nodes—at 130, 90, and 65 nm—are sufficient to make the optical waveguides.

Compared to today's 14-nm CMOS processes, these larger CMOS sizes are archaic. CMOS manufacturing equipment already mothballed has been given a new lease of life, thanks to silicon photonics. And such processes, no longer state of the art, are far cheaper to operate. In turn, the optical masks used to pattern and construct the chips are a lot cheaper to make for these older processes.

In summary, while silicon photonics is driving its own requirements, being able to use a chip fabrication plant brings huge advantages such as precision manufacturing, device yield, and volume manufacturing, which ultimately promises cheaper chips.

Silicon photonics is not about using silicon for everything. That misses the point, says Professor Bowers. The key element is using silicon as a substrate—on 12-in. wafers rather than the smaller 2-,3- and 6-in. wafers used for indium phosphide or gallium arsenide optical devices—and having all the process capability of a modern silicon CMOS facility.

1.2.1 The Application of Silicon Photonics

There are several battle lines between the different technologies when it comes to communicating data. And these battle lines are shifting as data rates continue to grow.

One competitive battle is between optics and copper wire. Optical technology has long secured the role of sending core network traffic over long distances. The amounts of data are too vast and distances too great—hundreds and even thousands of kilometers—for this to be done using copper wire. That is because copper's capacity-reach product is orders of magnitude lower than that of optical fiber.

The current battleground between copper and optics is over distances of several meters to a few kilometers. Copper wire is still used to deliver data to the home in the form of telephone wires and broadband services, but when it comes to data centers and the higher gigabits-per-second links used to connect equipment, copper starts to run out of steam after a few meters. The underlying trend is that optics continues to advance, slowly pushing copper's use to ever shorter distances.

There is also competition within optics, between the three main technologies used to implement communications. Indium phosphide has secured the long reach while vertical-cavity surface-emitting lasers (VCSELs), made using gallium arsenide, are used for tens of meters to a few hundred meters. And like copper, the reach of VCSELs diminishes as data rates continue to rise.

Silicon photonics is the newest of the three optical technologies. Being silicon-based, the technology suits being used ever closer to electronic chips. Silicon photonics has also been shown to work within an electronic chip, sending and receiving data on and off the chip. Such an application of silicon photonics is still some way off commercially, but it is coming. Fig. 1.1 shows the breadth of silicon photonics' reach.

The issues of bandwidth and reach for the different optical technologies and for copper are discussed in Chapter 2, Layers and the Evolution of Communications Networks. Note that Fig. 1.1 segments

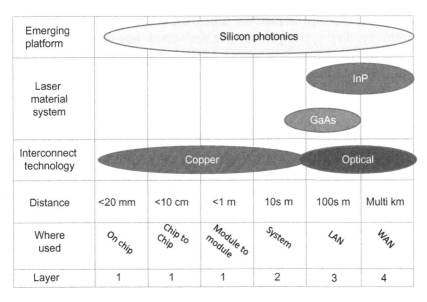

Figure 1.1 The reach of silicon photonics and other interconnect technologies. Based on "On-board Optical Interconnect," CTR III TWG Report #3, MIT Microphotonics Center, April 2013, figure 2.

distances into layers, a classification we introduce and expand upon in Chapter 2, Layers and the Evolution of Communications Networks.

1.3 THE SIGNIFICANCE OF SILICON PHOTONICS

The central tenet of silicon photonics is that it has an edge based on its ability to exploit the huge investment made over decades in the mass production of semiconductor chips. This is not something that established optical technologies such as indium phosphide and gallium arsenide can benefit from to the same degree. But while the basic premise is sound, the early reality of silicon photonics has proved more complex.

Silicon photonics uses silicon-on-insulator wafers. These 12-in. wafers are not mainstream silicon wafers but ones developed specifically for high-speed electronic circuits. However, silicon photonics benefits from the wafer's insulating layer for optical waveguide construction. Optics uses three styles of paths to guide light. One is optical fiber, another is a planar waveguide, and the third is free-space optics. Optical waveguides play a central role in all the main silicon photonic circuit building blocks.

Much investment has been made—and more is needed—for silicon photonics to exploit the manufacturing capabilities of the semiconductor industry. So far, companies have developed their own particular silicon photonic devices that require different manufacturing processes, integration approaches, and schemes to attach the laser to the chip. As mentioned, silicon does not lase.

Another issue is component volumes. The success of the chip industry stems from its huge volumes. In 2015 some 1.4 billion smartphones [3] and 276 million laptops [4] were sold. In contrast, Acacia Communications—an early success story of the silicon photonics industry—announced in 2016 that it had sold 13,000 of its silicon photonics−based coherent optical transceivers, and that was over 2 years [5]. This equates to a few hundred 12-in. wafers a year—a trifling volume for today's state-of-the-art semiconductor fabs. Such volumes hinder the full weight of the semiconductor industry coming into play and limit the interest of the big chip-making foundries—firms such as TSMC and GlobalFoundries that make the chips for "fabless" semiconductor firms.

So the notion of riding to prominence on the back of the chip industry has caveats, but silicon photonics has already benefited from the chip industry, and it will benefit the chip industry overall as is now explained.

To understand the significance of silicon photonics, several perspectives are used: what the semiconductor industry brings to silicon photonics, how silicon photonics benefits the chip industry, and how silicon photonics benefits systems design.

1.3.1 Silicon Photonics From a Photonics Perspective

From a photonics perspective, silicon photonics gains several inherent advantages by piggybacking on the chip industry.

- *Larger wafers*: Silicon photonics can exploit the chip industry's much larger 12-in. silicon wafers. Given that some photonic circuit designs can occupy a relatively large die area, the cost of silicon photonic chip-making is lower than competing technologies due to more devices being made simultaneously on a wafer—a basic premise of the chip industry.
- *Manufacturing precision and device yields*: The chip industry has invested trillions of dollars in the chip-making process using masks, photoresist materials, and etching to shrink continually the feature sizes of transistors. Using such advanced manufacturing results in device yields that are higher compared to that of traditional photonic integrated circuits made using indium phosphide. Indeed, higher yields are what got Acacia Communications' Chris Doerr, a veteran in indium phosphide photonic integrated circuit design, hooked on silicon photonics [6].
Doerr was at the famed US research institute, Bell Labs, for 17 years before joining Acacia in 2011. He spent the majority of his time at Bell Labs making indium phosphide−based optical devices and also planar lightwave circuits. Doerr had an opportunity to design a photonic integrated circuit using silicon photonics and chose to make a coherent receiver used for optical transport. What hooked Doerr was the high yield of the silicon photonic devices he made. He could assume each device worked, whereas when making complex indium phosphide designs, he would have to test half a dozen devices before finding a working one. And since yields were high, he could focus on making complex designs—devices with many photonic elements.

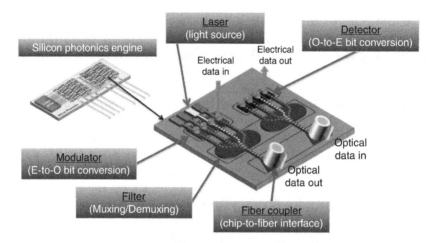

Figure 1.2 Silicon as an optical integration platform. Julien Happich, "CMOS-compatible intra-chip photonics brings new class of sensors", <http://www.analog-eetimes.com/news/cmos-compatible-intra-chip-photonics-brings-new-class-sensors>, Courtesy of EETimes, October 15, 2013.

Silicon photonic circuits also match closely their simulation results. Indium phosphide is complex, says Doerr, now Acacia's associate vice president of integrated photonics, and designers have to worry about composition effects and etching not being that precise. With silicon, the dimensions and optical performance are known with precision, meaning a design can also be simulated precisely, enhancing the design process.

- *Silicon as an integration platform*: Silicon not only implements optical waveguides and modulators (see Fig. 1.2)—used to encode digital data onto light before transmission—but also makes use of other materials such as germanium for photodetection. But it doesn't end there. It is also possible to bond III-V materials such as indium phosphide and lithium niobate, a mainstay material for optical modulation, onto silicon and process it to become part of the working photonics circuit. This is the best of both worlds: using III-V and other materials when needed and the larger wafers and the processing of silicon [7]. This topic is discussed further in Chapter 3, The Long March to a Silicon-Photonics Union.

1.3.2 Silicon Photonics From a Semiconductor Perspective

The semiconductor industry has benefited from Moore's law for over half a century. This has resulted in the continual advancement in the processing power of computer chips and ever denser memory chips, delivered like clockwork with each new generation of integrated circuit while costing about the same. Now the chip industry faces its own challenges as Moore's law finally comes to an end.

The industry is looking at novel ways to squeeze more life out of Moore's law but realizes that other strategies are needed if progress is to continue at the rate that Moore's law has consistently delivered.

One issue is getting data on and off the chip. For more advanced chips such as switch chips that are used to connect the computing resources in a data center, this input−output issue is becoming a pinch point. Optics can be used to tackle chip input−output bottlenecks as well as extend the reach while lowering the power consumption compared to today's high-speed electronic signaling (Fig. 1.3).

Electronics already plays an important role for optics. It can be used to improve the performance of optics, enabling what some call smart photonics. Examples include the digital signal processing techniques used to compensate optical transmission impairments in fiber over long distances, detailed in Chapter 5, Metro and Long-Haul Network Growth Demands Exponential Progress, as well as circuitry to control the stability of certain optical devices for tasks such as modulation.

Indeed, optics and electronics have even been combined on a single silicon photonics chip. Two companies, Luxtera and IBM, have developed single-chip monolithic designs combining the drive electronics with optics. Luxtera and the industry in general have since chosen to separate the two domains, with each on a separate chip fabricated using an appropriate CMOS process node: an advanced feature node

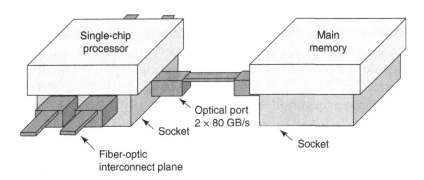

Figure 1.3 Optics to interconnect chips. © 2016 IEEE. Reprinted, with permission, from Levi AFJ. Challenges for photonics in future systems. In: Biophotonics/optical interconnects and VLSI photonics/WBM microcavities, 2004 Digest of the LEOS summer topical meetings; 2004. p. 1−2. http://dx.doi.org/10.1109/LEOSST.2004.1338675.

for the electronics and an older CMOS node suited for the larger optical features. The two chips are then co-packaged.

Creating compact single-chip photonic devices and using electronics to improve the performance of optical designs highlight the beneficial relationship between photonics and electronics. But the most significant development is how silicon photonics promises to benefit chip design in terms of greatly enhancing a chip's input−output capacity. And by doing so, it also benefits systems design, as discussed in Chapter 7, Data Center Architectures and Opportunities for Silicon Photonics.

1.3.3 Silicon Photonics From a System Perspective

When we refer to a system, we mean equipment that combines several technologies—or functional blocks—to perform a task. This task typically involves taking an input and processing it to produce an output. One example is a server, a rack of cards on which sit advanced microprocessors, memory, storage, and networking chips. The systems encountered in this book comprise various technologies to perform such tasks as computing, networking, data storage, and sensing, or combinations of these tasks.

One trend in the data center is the development of disaggregating systems, where the basic elements that make up a system are being rearranged and separated in a nontraditional way to deliver a performance or an operational benefit. But for this to work, what is needed is a high-bandwidth, cheap interconnect to link the disparate parts of the system.

One example is the disaggregated server. Here, the components making up the server—the processors, storage, and memory—are pooled separately but interconnected using high-speed links (Fig. 1.4).

The benefits of a disaggregated architecture are that different units can be upgraded as required—e.g., a processor node—without having to upgrade the complete server, thereby saving expense. Equally, the cooling needed for the equipment can be customized to the particular units generating the most heat.

At present such links in disaggregated systems can be made with copper or with VCSELs, but as the system elements continue to improve in performance and speed, this is an obvious application for silicon photonics. We discuss this in Chapter 7, Data Center Architectures and Opportunities for Silicon Photonics.

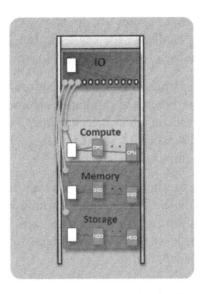

Figure 1.4 Disaggregated server where compute, storage, and memory are each pooled and they share networking.
Max Smolaks, "Ericsson to sell Intel's hyperscale kit to network operators", <http://www.datacenterdynamics. com/content-tracks/servers-storage/ericsson-to-sell-intels-hyperscale-kit-to-network-operators/93484.fullarticle>, March 3, 2015.*

Another example of how silicon photonics can benefit systems is a microprocessor design that uses optics for its off-chip communications. A group of academics from several US universities have used a standard 45-nm CMOS process from IBM to create an advanced processor having optical input–output on a single chip [8]. Here the bulk of the chip area is digital logic, with the silicon photonics accounting for a small fraction of the die area. Such a design is not yet commercially viable, but it is a key milestone because it already demonstrates a capability that will be needed in future.

These systems examples highlight the beneficial relationship between photonics and electronics, with photonics delivering bandwidth and performance benefits.

Indeed, many argue that for the chip industry to progress, greater thought must be given to system design innovation, especially as Moore's law, with its guaranteed chip performance benefits delivered

*Every effort has been made to trace copyright holders and to obtain their permission for the use of copyright material. The publisher apologizes for any errors or omissions in the acknowledgements printed in this book and would be grateful if notified of any corrections that should be incorporated in future reprints or editions.

on cue every two years, comes to an end. System innovation has always been a design challenge for engineers but promises to be even more so in future because of the demise of Moore's law.

In recent years it has been the systems vendors—companies such as Cisco Systems, Juniper Networks, Mellanox, Huawei, and Ciena—that have been acquiring silicon photonics startups. They realize they will need silicon photonics expertise for the design of their own systems and for their custom chips.

Silicon photonics can therefore be seen as an important and timely technology adjunct for the chip industry. Andrew Rickman, tackling switch system innovation for the data center with his startup Rockley Photonics, is a firm believer that system design is where silicon photonics promises unique benefits [9].

For Rickman, using silicon photonics to make an optical component is not playing to its strengths. You can look at nanoelectronics in isolation and use silicon photonics for chip-to-chip communications, a good thing to do, he stresses. Or you can address, as Rockley aims to do, Moore's law and the input−output limitations within a complete system the size of a data center, where hundreds of thousands of computers are interconnected. Clearly the scale of the problem being tackled—data center scaling versus solving an individual problem around a chip—results in a different approach, says Rickman.

Professor Richard Soref, described as the founding father of silicon photonics, holds a similar view. He is already thinking about new emerging applications for silicon photonics such as sensors and microwave photonics [10].

Soref talks about the near-infrared and part of the mid-infrared spectral range: 1.5−5 μm. This is above the spectral range of light used for datacom and telecom applications. VCSELs operate at 0.85 μm typically while long-range optical transport is around 1.5 μm.

The applications in this higher spectral range include system-on-a-chip, lab-on-a-chip, sensor-on-a-chip, and sensor-fusion-on-a-chip for such applications as chemical, biological, medical, and environmental sensing. Such sensor chips could find their way into your future smartphone and play an important role in the emerging Internet of Things, says Soref.

These are different systems from the ones associated with telecom and datacom but systems nonetheless.

1.4 THE STATUS OF SILICON PHOTONICS

Silicon photonics has still to reach its tipping point despite being used in commercial products. By tipping point we mean the point when the technology transitions from early specialist applications to become pervasive.

The concept of a tipping point, popularized in the bestseller by Malcolm Gladwell, describes the moment when an idea, trend, or social behavior crosses a threshold and spreads like wildfire [11].

Clearly silicon photonics is still in its infancy and cannot be described as pervasive. And technology always advances at a more sedate pace than wildfire. But while it took close to 20 years (1985−2005) for silicon photonics to advance from a core idea to first products, the last decade (2005−15) has seen significant progress in its development. Now, various elements are being put into place such that silicon photonics is entering the final straight, with the tipping point finishing line within sight (see Fig. 1.5).

Luminaries that have done much to develop the field of silicon photonics have conflicting views as to whether the technology has reached its tipping point.

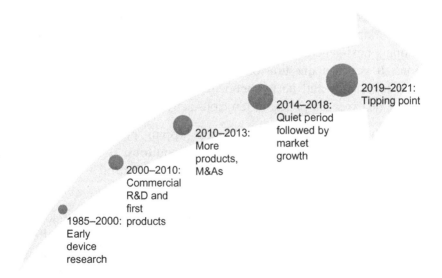

Figure 1.5 Silicon photonics' evolution from a core idea in 1985 to approaching its tipping point.

Mario Paniccia, who headed Intel's silicon photonics program until 2015, defines silicon photonics' tipping point as when people start believing the technology is viable and are willing to invest [2]. AIM Photonics, the $610 million publicly and privately funded initiative set up in 2015, is one indicator of that willingness [12]; Acacia's successful initial public offering in May 2016 is another. Not only is silicon photonics being seen as viable, but players are investing significant funds to commercialize the technology. Paniccia also points to the many companies now selling commercialized silicon photonics. In his view, silicon photonics' tipping point has passed.

Another luminary, Graham Reed, Professor of silicon photonics at the University of Southampton's Optoelectronics Research Centre, gives an academic perspective. In the 1990s it was a challenge to get funding to research silicon photonics, he says. Now, his group is regularly approached by companies from all over the world that are either active in silicon photonics or plan to enter the market.

Given how some of the largest companies have been investing in silicon photonics for over a decade, Reed is surprised that more products are not available commercially. But he believes it is inconceivable that the firms that have made the investments will not launch products. So, in that sense, the tipping point has already been and gone, argues Reed. All these events are of a technology pointing to mass market, he says.

Reed describes this as an industry "quiet period," with photonics companies working to commercialize their technologies, and systems vendors developing next-generation products and evaluating various technological options. It is just a question of time before a vendor "jumps" and the market takes off—and he is confident somebody will jump. Once that happens, he expects a period of ferocious activity to follow [13].

Other luminaries do not yet see the tipping point, Richard Soref being one. For silicon photonics to be ubiquitous, it will take much investment and many commercial results, he says, and we are not at that stage.

There are hurdles for silicon photonics to overcome, and some of them are chicken-and-egg ones. Foundries will be interested in offering silicon photonics fabrication services once there are large unit volumes, but that requires a variety of companies—fabless chip players, sensor companies, and systems houses—to have their designs made in volume.

At present companies are developing custom solutions for manufacturing and packaging, as mentioned. This presents a barrier to entry to those interested in using the technology but lacking the resources to develop their own solutions. And while some of the existing silicon photonics companies may consider making their technology available to the industry, they will only make additional investment if there is enough interest—another chicken-and-egg situation.

A key industry focus is to improve the manufacturing of silicon photonics circuits on a commercial scale, a hurdle that must be overcome for widespread use to occur. Silicon photonics players such as Acacia Communications, IBM, Intel, Luxtera, Mellanox Technologies, STMicroelectronics, Macom, Kaiam, Cisco Systems, Oracle, and Juniper Networks are investing and developing silicon photonics circuit design and packaging, but they are using custom approaches based on their own intellectual property. Until there are higher volumes, foundries may not make the investment to take their intellectual property to the next level to offer it commercially to interested parties.

There is a complex interaction between technology advancement, volume demand, and manufacturing progress, as shown in Fig. 1.6. The technical and manufacturing challenges facing the industry are engineering issues rather than show-stoppers. Engineering effort and investment will thus resolve these challenges rather than requiring some

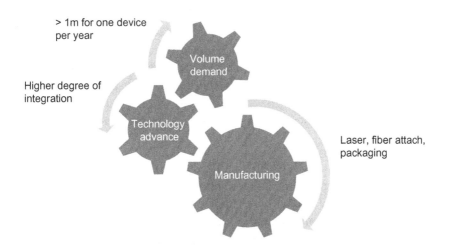

Figure 1.6 The remaining stages before silicon photonics reaches its tipping point.

fundamental innovation that cannot be scheduled or predicted. This is why we are confident silicon photonics will reach its tipping point.

Many of the conditions that must be met for silicon photonics to reach its tipping point are now in place. Investment in the technology is healthy and increasing, more companies are embracing it, and a wider base of products is coming to market.

But the technology is not widely adopted, and applications that will deliver significant unit volumes have yet to start.

We expect silicon photonics to break out in the coming years. At some stage around 2021, it will become evident that silicon photonics has reached its tipping point. What is happening now can best be described as entering the era of silicon photonics.

1.5 SILICON PHOTONICS: MARKET OPPORTUNITIES AND INDUSTRY DISRUPTION

For completeness, we end this chapter by touching briefly on two other core issues regarding silicon photonics aside from its significance and status. The first is the market opportunities going forward for the technology. There is also the question of whether silicon photonics is disruptive. These two issues are discussed in Chapter 8, The Likely Course of Silicon Photonics, which brings the book to a close.

1.5.1 The Market Opportunities for Silicon Photonics

Optical transport for telecom and connecting equipment in the data center are applications that are already using silicon photonics commercially, and these will be the main markets driving the technology in the near term—from now till 2021. Chapter 8, The Likely Course of Silicon Photonics, presents silicon photonics market research forecast data supporting this.

Mid-term, from 2021 to 2026, the datacom and telecom markets will continue to grow, with silicon photonics being used for shorter-reach links currently served by VCSELs. During this mid-term period, other markets—sensors, devices for microwave photonics, medical devices, Light Detection and Ranging (LIDAR, a technology similar to radar that uses laser light to distinguish much finer details of an environment and which can be used in vehicles and robotics), and non-traditional computing—will begin to emerge.

What about pitfalls? If silicon photonics takes too long to develop, companies pursuing the technology could struggle to generate sufficient revenues and it could end up being a niche technology.

We do not believe that will happen. While traditional photonics technologies are proven, silicon photonics has unique advantages. Investment in silicon photonics is growing, as is the number of players pursuing the technology. And the required supply chain needed by companies to exploit the technology is taking shape.

The visible market of optical technologies linking equipment in the data center and across telecom networks may be limited, but system vendors will increasingly use silicon photonics in their equipment. Mid-term, this trend will continue: silicon photonics will be used for more and more interconnect applications, boosting volumes further. And the technology will find its way into systems such as servers and switch architectures to address the scaling of data center networking and the linking of data centers. Overall, market size will thus grow in design variety and in revenues.

And newer markets will also emerge. Their timing and prospects are less clear, and the pace of adoption of silicon photonics in these markets remains speculative. But these markets have requirements that will benefit from silicon photonics and its growing maturity fostered by the telecom and datacom markets. These markets will also drive their own requirements, advancing the technology and refining its manufacturing processes.

1.5.2 Silicon Photonics is a Disruptive Technology

Silicon photonics does not meet the classical definition of a disruptive technology—one that arrives meeting the low end of a market's product requirements—yet it has the attributes to be disruptive. The technology is improving rapidly. And companies will be able to design a photonic circuit and have it made in a foundry. This "fabless" model is well known in the chip industry but is new to the optical component industry. Thus silicon photonics promises to change the traditional optical vendor supply chain and thereby disrupt the optical marketplace.

We equate the impact of silicon photonics to tectonic shifts and earthquakes. Tectonic shifts occur, but their effect in causing an earthquake is not immediate. Silicon photonics is the tectonic shift of the optical world, but its full impact has yet to be played out.

Finisar, the world's leading optical component company and expert in indium phosphide and VCSEL technologies, has started working with silicon photonics for good reason. Notwithstanding the company's earlier statements expressing reservations about silicon photonics [14], it has embraced the technology because it feels the plates shifting.

But perhaps the biggest disruption will result from having the technology embraced by leading chip companies. This clearly has started: Intel is one, STMicroelectronics is another.

When the technology starts to be offered as part of the chip design environment rather than as a photonic technology embraced by the optical industry for their component designs, this will be a tipping-point catalyst and disruptor combined.

Key Takeaways

- The advent of the transistor has given rise to ubiquitous computing, while optical technology has enabled the global transfer of data and the Internet. Silicon photonics brings together the two technologies—electronics and optics—to enable the continuing evolution of the digital economy.
- Silicon photonics involves the making of photonic devices on a silicon substrate fabricated in a chip CMOS fabrication facility.
- Interconnect technologies competing with silicon photonics include copper and established optical technologies such as indium phosphide and gallium arsenide. Silicon photonics can be used across a vast scale of distances for interconnect: within and between chips all the way to wide area networks spanning thousands of kilometers.
- Silicon photonics is more than just an interconnect technology. It will benefit chip design and the input–output bottleneck. It also benefits systems design, especially important with the demise of Moore's law.
- The tipping point for silicon photonics has not been reached, but the main elements needed for the technology to become pervasive—including volume manufacturing—are being put in place.
- The main market applications for silicon photonics are telecom and datacom in the near term. Newer markets will also emerge such as sensors, LIDAR, medical devices, and microwave photonics.
- The disruptive force of silicon photonics will become evident when the semiconductor industry offers optics as a design element in its toolbox.

REFERENCES

[1] Optical integration and silicon photonics: a view to 2021. Gazettabyte, < http://www.gazettabyte.com/home/2016/5/12/optical-integration-and-silicon-photonics-a-view-to-2021.html >; May 12, 2016.

[2] Mario Paniccia: We are just at the beginning. Silicon photonics luminaries series. Gazettabyte, < http://www.gazettabyte.com/home/2016/5/23/mario-paniccia-we-are-just-at-the-beginning.html >; May 23, 2016.

[3] Gartner says worldwide smartphone sales grew 9.7 percent in fourth quarter of 2015, press release, < http://www.gartner.com/newsroom/id/3215217 >; February 18, 2016.

[4] PC Market Finishes 2015 as expected, hopefully setting the stage for a more stable future, IDC, press release, < https://www.idc.com/getdoc.jsp?containerId = prUS40909316 >; January 12, 2016.

[5] Acacia Communications ships over 13,000 of its coherent CFP modules, press release. < http://ir.acacia-inc.com/phoenix.zhtml?c = 254242&p = irol-newsArticle&ID = 2164387 >; March 18, 2016.

[6] Enabling coherent optics down to 2 km short-reach links. Silicon photonics luminaries series, Chris Doerr. Gazettabyte, < http://www.gazettabyte.com/home/2016/6/27/enabling-coherent-optics-down-to-2km-short-reach-links.html >; June 27, 2016.

[7] Heterogeneous integration comes of age. Silicon photonics lumixnaries series, John Bowers. Gazettabyte, < http://www.gazettabyte.com/home/2016/8/28/heterogeneous-integration-comes-of-age.html >; August 28, 2016.

[8] Su C, et al. Nature December 23, 2015;528(7583):534−8.

[9] Tackling system design on a data centre scale., Silicon photonics luminaries series, Andrew Rickman. Gazettabyte, < http://www.gazettabyte.com/home/2016/5/6/tackling-system-design-on-a-data-centre-scale.html >; May 6, 2016.

[10] Richard Soref, The new frontiers of silicon photonics. Silicon photonics luminaries series. Gazettabyte, < http://www.gazettabyte.com/home/2016/6/17/richard-soref-the-new-frontiers-of-silicon-photonics.html >; June 17, 2016.

[11] Gladwell M. The tipping point: how little things can make a big difference. Boston: Little, Brown and Company; 2000. p. 288.

[12] Making integrated optics available to all. Optical Connections, Issue 7, Q3 2016, p16.

[13] Professor Graham Reed: The calm before the storm. Silicon photonics luminaries series. Gazettabyte, < http://www.gazettabyte.com/home/2016/5/28/professor-graham-reed-the-calm-before-the-storm.html >; May 28, 2016.

[14] Q&A with Jerry Rawls, Part 2. Gazettabyte, < http://www.gazettabyte.com/home/2013/9/17/qa-with-jerry-rawls-part-2.html >; September 17, 2013.

CHAPTER *2*

Layers and the Evolution of Communications Networks

Layers seems to be everywhere in mother nature.

Michael S. Gazzaniga [1]

Everything that has happened in the telecom network is now being replicated inside the data center. And then everything that is happening in the data center is going to be on the board, and then everything on the board is going to be in the package, and then everything in the package is going to be on the chip.

Lionel Kimerling [2]

2.1 INTRODUCTION

I [Rubenstein] have my father's slide rule on my desk. It sits in a dark green box whose worn corners hint at years of use.

For readers familiar only with the digital age: the slide rule was a constant companion of engineers and academics. An ingenious mathematical device, it was used to multiply, divide, calculate roots, and work out logarithms by aligning numbered rulers against each other with the aid of a movable glass window inscribed with a line.

Fast-forward half a century and the WolframAlpha application runs on my tablet computer. I tap in numbers or a function, press a key, and the answer appears in multiple forms: numerically, as functions, and graphically, plotted in several ways.

The tablet does not perform the calculations, although you would not know, given how quickly the answers appear. And the tablet itself is no mathematical slouch. It is an extremely powerful computer with a microprocessor chip made up of two ARM 64-bit cores, each working to a clock beating over a billion times a second.

Silicon Photonics. DOI: http://dx.doi.org/10.1016/B978-0-12-802975-6.00002-8
Copyright © 2017 Daryl Inniss and Roy Rubenstein. Published by Elsevier Inc. All rights reserved.

Instead, the calculation is sent, via the local service provider, through the Internet to find WolframAlpha's computers—servers—housed in a data center. The servers do the calculations and the results are returned to the sender. How do I know? Well, if I disconnect Wi-Fi and go offline, a polite screen message appears: "Sorry, you need an Internet connection to do that." Another reason to keep the slide rule close at hand.

If the tablet is a powerful computer, why send the calculation to some remote location hosting computers and wait for the answer to return? The glib answer is that the network is now sufficiently advanced to enable this. Using optical networking, data can be sent, computed, and returned in a fraction of a second. Here, data in the form of an electrical signal is converted into light and sent over an optical fiber before being restored as an electrical signal at the servers. The same signal translation process is undertaken when the data is returned to the tablet. And WolframAlpha offers more knowledge resources than just math.

2.1.1 The Cloud as an Information Resource

This model of a "dumb terminal" talking to centralized computing is not new. An early example was the mainframe computer where requests were sent via a terminal into a central computing room located in the same building or campus. Now, networking allows for the computing to be hundreds or even thousands of kilometers away from the user, while the computing resources—based on tens of thousands of computers linked in a data center and even between data centers—bear little resemblance to the early mainframes.

Nearly all the operations you undertake online involve interactions with remote computers or servers. The name *server* reflects how the computer unit "serves up" data and remote applications. Through such user requests—e.g., Internet searches or hitting a "Like" button—a wealth of data is accumulated, including your habits, preferences, and interests.

Such data fuels the businesses of Amazon, Google, and Facebook—referred to as Internet content providers—and explains the rapid rise of new industries such as Big Data. Companies in many segments are building data centers to host and "mine" this treasure trove of information. Car manufacturers, e.g., want data not only about the state of your connected vehicle but how you are driving it, your routes, and where you stop for coffee. In other words, they want to know your life-style—and that of everyone else driving their vehicles [3].

The cloud, as described, is characterized largely by human-triggered transactions. But soon it will be common for machines to be connected to the Internet, a development known as the Internet of Things.

Imagine an airborne drone navigating its way to deliver a parcel. For a successful delivery, the autonomous drone will need to navigate its way to the destination, requiring reliable networking and rapid interaction with remote computing. And the drone will be one of many in the sky.

Estimates suggest that Internet-connected machines could number between 20 and 30 billion as early as 2020 [4]. The amount of data generated, transmitted, and analyzed will accelerate as more and more items become connected, placing new demands on data centers and the network.

Having cloud-based centralized IT resources is also changing the way enterprises work. Instead of having their own engineering staff to purchase and maintain hardware and software, cloud providers can do it for them. By centralizing the task for multiple enterprises, the cloud providers can achieve economies of scale that only the largest of enterprises can match.

2.1.2 Optical Networking is Central to the Internet

Two huge industries, telecommunications and data communications, are required for the WolframAlpha application to work. Communications service providers—telcos—operate the networking infrastructure that underpins the Internet, while the Internet businesses create the clever services like the tablet application using cloud computing that run on top of the telecom networks.

Within the data center, cabling and networking switches are used to connect the IT equipment such as servers and storage (Fig. 2.1). The networking also includes Internet Protocol routers and optical trans-mission equipment to connect the data center to another data center or to the general network, referred to as the wide area network or WAN.

Some of the largest data center operators own optical fiber and oper-ate networking equipment, but the telcos are required for some part of the data's journey between users and the information they receive.

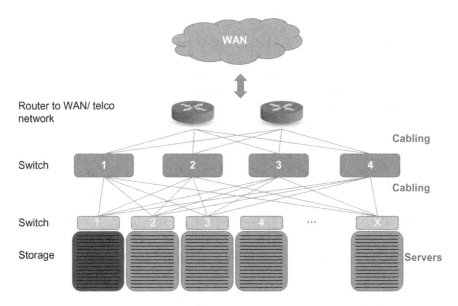

Figure 2.1 Data center networking schematic showing servers, routers, and storage networked with cables and switches.

2.1.3 Silicon Photonics: A Technology to Tackle the Industry's Challenges

Both the telecom and datacom industries have pressing requirements that are stretching technologies to the limit. This book concentrates on the business and technology trends, explains the growing role of optical technology in networks, and discusses why silicon photonics is well-positioned to deliver solutions. Moreover, datacom and telecom are just two markets for silicon photonics, albeit two very important ones.

Before reflecting on the industry challenges, it is important to look at telecom and datacom more closely by segmenting these two increasingly intertwined worlds into layers. By highlighting the role of optics at each of these layers, the emerging market opportunities for silicon photonics become clearer.

2.2 THE CONCEPT OF LAYERING

Cells, bacteria, and the brain are all examples of complex systems in nature, and they all share a layered structure. Layering turns out to be a useful tool to analyze complex systems, whether natural or man-made [1]. Telecom and datacom combined can be viewed as one such complex system.

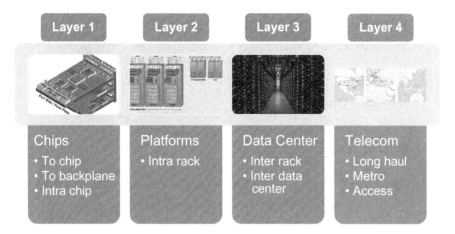

Figure 2.2 The four-layered model for cloud and telecoms: chips, platforms, the data center, and telecommunications networks. From: Layer 1 Synopsys, Layer 2 Cisco, Layer 3 Google, and Layer 4 GTT Communications.

Fig. 2.2 shows a four-layered model of cloud and telecoms. What becomes clear is that the layers share common characteristics. For example, the three pillars of the information age—communications, computing, and storage—are found at each of the four layers. What distinguishes the layers is their scale or, more accurately, their reach—an important metric in communications.

Each layer can be viewed distinctly, with its own implementations and rules. This is a common characteristic of layered systems and explains the power of the approach: breaking down the system into its layered elements and then analyzing each layer allows a complex system to be simplified. In turn, details can be encapsulated and abstracted at the individual layers.

The layers, now described, start with Layer 4 telecom networks and drill down to the shortest reach, Layer 1: the chip level.

2.3 THE TELECOM NETWORK—LAYER 4

The telecom network is the topmost layer, Layer 4. This is not one network but a collection of networks, run and managed by the telcos. The networks range from the link between your home and the local exchange to global communication links that carry traffic across continents and even between continents using submarine cables (Fig. 2.3).

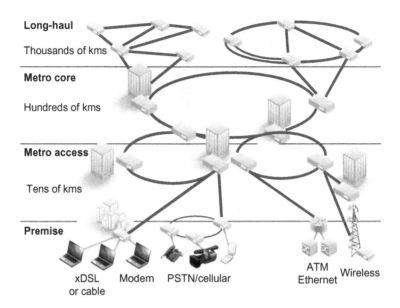

Figure 2.3 A telco network including long-haul, metro core, metro access, and premises, and their typical reaches.
Exfo. "From EXFO, http://www.exfo.com/solutions/metro-core-networks/bu2-bu3-packet-optical-transport/
technology-overview".

Telcos may also provide cloud-based services, but for this discussion we focus on their role in enabling the cloud's workings in general by overseeing the complex networks that underpin the Internet.

Telcos' networks serve large enterprises and billions of individual users. The telcos also offer wholesale services to other operators. Their businesses generate huge revenues: the world's 15 largest telecom operators have annual revenues that, when combined, exceed one trillion (one thousand billion) US dollars, while the total telecom operator market is worth nearly $2 trillion annually.

The networks in Layer 4 can be categorized into three types: long-haul, metropolitan (metro) and regional, and access.

- *Long-haul networks* span distances from 1000 to as many as 13000 km in the case of pan-Pacific submarine cabling. The networks use dense wavelength-division multiplexing, an important optical technology first used in the network in the 1990s. Dense wavelength-division multiplexing sends light at different wavelengths or colors down an optical fiber. Each color is like a unique traffic lane. With modern optical transmission technology, the highways carrying this traffic are often designed with 96 wide lanes, each one carrying 100 Gb—100 billion bits—or even 200 Gb of data each second. Since optical transmission usually sends the data in one direction down a fiber, a fiber pair is needed to send traffic in both directions.

- *Metro and regional networks* serve a city or a region, respectively. Distances here range from 40 to 1000 km, with dense wavelength-division multiplexing the predominant technology used. These networks are important for cloud computing because they link data centers sprinkled across a metropolitan area. Data center interconnect is growing in importance, but it is just one traffic type of many carried by metro and regional networks.

- The *access network* refers to the edge of the network, closest to the user or an enterprise. Distances here span to 40 km, although most access links are shorter. The radio access network—the radio part of the mobile network linking the cellular mast and your handset—and the fixed or wired network that brings broadband to your neighborhood and home are located here. Radio technologies, copper wires such as the legacy telephone network, and optical broadband technologies like fiber-to-the-home are used in the access network.

Telcos also operate aggregation points—buildings that house lots of equipment—in their networks, known as points-of-presence. They also operate central offices where equipment that aggregates traffic—e.g., radio and broadband data—reside. And as the operators increasingly embrace servers to deliver services, these buildings are now also housing data center equipment.

2.3.1 Cloud Computing is Driving the Need for High-Bandwidth Links Between Data Centers

The linking of data centers owned by the large Internet companies has become an important subcategory of optical networking. One reason is the rapid construction of data centers worldwide and the central role they play for Internet businesses, a topic addressed in Chapter 6, The Data Center: A Central Cog in the Digital Economy. These players are some of the most profitable companies in the world. Another reason why linking data centers is important is the significant and growing traffic they need to transport, which is leading to new equipment requirements; this is addressed in Chapter 5, Metro and Long-Haul Network Growth Demands Exponential Progress.

Statistics shared by two Internet giants reveal how the worlds of telecom and datacom have become intertwined.

Google has stated that a single Internet search query travels on average 2400 km before it is resolved [5]. The huge resources used by, and the sophistication of, Internet search algorithms are well documented. But what is revealing is how resolving a query can involve distributed data centers. Moreover, no matter how advanced Google's data center architecture and search algorithms are; dense wavelength-division multiplexing communications must play a role given the distances involved.

A second example comes from Facebook, where a single user request (a hypertext transfer protocol or http request) to one of the company's web servers generates a near 1000-fold increase in internal server-to-server traffic [6]. What this highlights is how much internal data center traffic is generated from a simple transfer across the wide area network. Clearly, the amount of traffic in and between a company's data centers is far greater than the flow of traffic from the wide area network to and from the data centers. Indeed, estimates suggest that three quarters of data center traffic remains inside the data center [7].

Data centers can be distributed across several buildings in the same industrial park (access network distances), across a metropolitan area (metro distances), or, as we have seen, across countries or continents (long-haul distances). The amount of capacity needed to link data centers between local buildings or across a metro is huge: tens and even hundreds of terabits of capacity. A terabit is 1000 Gb. Links between data centers over long-haul distances are more akin to traditional long-haul dense wavelength-division multiplexing traffic. But unlike a telco's more general traffic, these tend to be point-to-point links.

2.3.2 The Importance of Optics for Communications

The discussion here has focused on Layer 4 as a communications layer, but storage and computing are also required. Storage is used, e.g., in IP routers to buffer the incoming IP packet streams until they are processed and routed. The network also needs to be controlled, and that requires software running on general-purpose microprocessors used in telecom platforms and, increasingly as telcos adopt datacom practices, on servers too.

From a communications perspective, optics is fundamental in enabling telecom networking. Electrical cables do not have the same reach as fiber, nor can they match fiber's information-carrying capacity, i.e., its bandwidth. Telecom started out over a century ago sending telephone calls over copper wire as analog signals. Aggregating all the calls coupled with other services running on the network soon required large-capacity pipes, and this accelerated further with the advent of digital data and services, causing telcos to turn to optical technology to send the summed traffic over long distances. Now phone calls— even high-definition voice calls—are but a trickle in the deluge of digital data types sent across the global telecom network.

Optical technology is the foundational layer on which the telecom network is constructed. And whenever optics is involved, so exists a market opportunity for silicon photonics.

2.4 THE DATA CENTER—LAYER 3

The data center is a cavernous warehouse crammed with information technology equipment. Walk inside one and you will see rows and rows of racks, each hosting tens of servers stacked on top of each other (Fig. 2.4). The largest data centers can host 100,000–200,000 servers.

Other key IT equipment in the data center includes storage and networking. Storage is needed to hold the data processed by the servers, while networking connects the storage with the servers. Because workloads require huge computational resources—servers organized in clusters— how the data center's equipment is networked is key to its processing performance, efficiency, and overall costs. This is the subject of Chapter 7, Data Center Architectures and Opportunities for Silicon Photonics.

Data centers have several important metrics. One is size, measured in terms of the floor space a data center occupies. The largest Internet businesses run huge data centers that can occupy more than 1 million square feet of floor area, and are referred to as mega- or hyperscale data centers.

Another important data center measure is power consumption. This includes the power consumed by all the equipment plus the additional power needed for the cooling systems to extract the heat generated by the equipment. The goal is to maximize the power efficiency of equipment while minimizing the power consumed for cooling. The issue of

Figure 2.4 Equipment inside a Facebook data center. © Copyright Facebook & Steve Tague Studios.

power consumption in the data center is also addressed in Chapter 6, The Data Center: A Central Cog in the Digital Economy.

A further distinction associated with data centers is the type of company operating them. The data center is at the core of the Internet business players such as Amazon, Facebook, Google, and Microsoft. These companies architect their data centers to maximize the efficiency of their operations. But since their operations differ, so do their requirements. IT is fundamental to all their businesses, though, and each introduced efficiency directly impacts their bottom line.

Large enterprises typically run smaller-sized data centers. Such data centers are clearly important for these companies' operations but are not their central businesses. As such, they will not employ as many data center staff as the Internet businesses, nor will they be at the forefront of driving technological progress. Indeed, these enterprises will be the main beneficiaries once the technological advances driven by the Internet businesses become mainstream.

2.4.1 Data Center Networking is a Key Opportunity for Silicon Photonics

The data center is one of the most dynamic arenas driving technological innovation, whether in servers, storage, networking, or software.

As such the data center represents a key market opportunity for silicon photonics. Server racks continue to evolve and need faster networking between them. And while copper cabling plays an important role in linking equipment in the data center—what we refer to as Layer 3—optical interconnect is gaining in importance as the amount of traffic rattling around the data center grows.

Copper cabling can link equipment up to 10 m apart but it is bulky, which can restrict equipment air flow and impede cooling. Copper cabling is also heavy: the sheer weight of connections on the front panel of servers or switches has been known to cause disconnects to equipment.

Optical connections are lighter, less bulky, and cover much greater distances—500 m, 2 km, and 10 km—more than enough to span the largest data centers.

Such connections are situated on the front of the equipment, referred to as the faceplate. The faceplate typically supports a mix of interfaces and technologies: electrical interfaces using copper cabling as well as optical links using fiber connectors and optical modules. Optical modules are units that plug into the faceplate. Modules support all sorts of speeds and distances within the data center (see Appendix 1: Optical Communications Primer). Speeds include 1, 10, 40, and 100 Gb/s; copper links may range up to a few meters, while optical spans up to 10 km. The optical modules, pluggable or fixed, can also use dense wavelength-division multiplexing to enable Layer 4 metro and long-haul distances. Such optical connections have become an early and obvious market for the silicon photonics players to target.

2.5 PLATFORMS—LAYER 2

The next layer down, Layer 2, is the equipment itself. For servers, these are typically boxes or rack units containing a printed circuit board on which sits the chips and optics. Rack units can be stacked in a chassis, up to 2 m high (Fig. 2.5). Data center managers can populate the chassis with as many racks as required, while individual racks can be changed without affecting the platform as a whole.

We define Layer 2 as communications within equipment—between boards or between rack units in a chassis. Typical distances are a few meters.

Figure 2.5 Cisco Systems' Nexus 7700 switch rack. From Cisco.

Telecom equipment is built differently from the IT equipment used in the data center. Telecom systems have slots across a shelf, each slot housing a printed circuit board vertically. The platform can be made up of one shelf or several stacked shelves. Telecom equipment is also built to a more stringent standard and must undergo lengthy qualification before deployment. This is because telecom equipment may be deployed in the network for 20 years or more, whereas equipment in the data center may be replaced every 2–3 years typically.

The backplane of a platform refers to the internal bus that connects the cards within the chassis. There has been repeated talk of systems moving from an electrical to an optical-based backplane, and such platforms have started to emerge commercially [8]. Oracle is one vendor that has announced switches—used to connect servers in the data center—that have an optical backplane. Another company is Ericsson, which has developed a server that uses optics to connect various elements making up the system. Such a design is known as a disaggregated server and is discussed in Chapter 7, Data Center Architectures and Opportunities for Silicon Photonics.

But Oracle and Ericsson are trailblazers; electrical signaling continues to be the dominant approach. Backplane electrical signaling is already at 28 Gb/s, and work is underway to double the rate to 56 Gb/s. The distances of electrical interfaces for backplanes at such high speeds are up to a meter.

Needless to say, silicon photonics' opportunities at Layer 2 include all applications of optics for communications within a system, including the backplane.

2.6 THE SILICON CHIP—LAYER 1

At the bottommost layer sits the silicon chip. Inside the chip, the internal signaling is sent across electrical busses to other digital logic building blocks. Microprocessors, the chips at the heart of servers, are made up of multiple central processing units or cores that are connected and operate in parallel, as well as memory (storage) blocks and peripheral circuits.

Examples of chips for telecom include the network processor, a specialized processor designed for processing traffic in the form of Ethernet frames and IP packets, and the coherent digital signal processor used for high-speed metro and long-distance connections. A digital signal processor is a chip designed to do lots of math—multiplies and adds—as quickly as possible. Layer 1 interconnect covers communications within chips, between chips on the same card, or between the chip and the faceplate (Fig. 2.6).

Layer 1 is probably the most exciting opportunity for silicon photonics—exciting in the sense that it promises larger volumes and because optics will bring to chips far more information-carrying capacity or bandwidth than electrical input−output. And that promises new styles of system designs.

Longer term still, the technology could even be used to connect the functional blocks within a chip. If the data center with its huge computing, storage, and interconnect resources can be seen as a hyperscale distributed computer, silicon photonics linking multiple processor cores

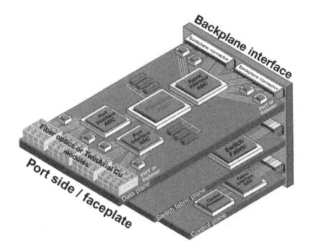

Figure 2.6 Line card showing connectivity to chips, faceplate, and backplane. Synopsys.

and memory within a chip can be viewed as performing a similar function, just on a scale two layers down. However, these are the shortest distances and copper has enough performance headroom to be the technology of choice for the foreseeable future.

The layered model representation of datacom and telecom highlights how optics are present at each of the four layers. What differs are the reaches, which diminish with each descending layer. Other differences include prices and unit volumes. Simply put, the use of optics over shorter distances equates to larger volumes overall, optics within chips being the extreme example of that.

2.7 TELECOM AND DATACOM INDUSTRY CHALLENGES

The leading US telco, AT&T, describes its role as providing fast, secure connectivity to everything on the Internet, independent of location and device [9]. This is an apt description for the role of the telecom industry in general.

The telcos provide mobile and fixed connectivity, and their networks carry IP traffic that continues to grow exponentially. Cisco Systems, a leading provider of IP core routers, forecasts that globally, IP traffic will nearly triple between 2015 and 2020, a compound annual growth rate of 22% [10].

With such rapid growth, it doesn't take many years before the consequences in terms of traffic on the network are felt. For telcos, traffic growth means having to invest continually in network infrastructure, addressing pinch points as they arise, whether in the long-haul, metro, or access networks that collectively make up Layer 4.

The telcos also operate in a competitive and regulated marketplace. The fierce competition comes not just from other telcos but from nimble Internet giants that deliver their services over the telcos' networks. The complaint often heard from telcos is that they have all the costs of running and investing in their networks while Internet businesses make lots of money with services that ride on top. Examples include streaming video service providers and messaging companies, known as over-the-top players.

Given telcos have annual revenues totaling $2 trillion dollars, not everyone is sympathetic. The *New York Times* in an editorial called this argument disingenuous. Telcos make money by charging customers monthly fees for accessing the network; if revenues were inadequate, operators would raise prices or become insolvent, it said [11].

A positive aspect of the competition is that the telcos are transforming their services and embracing practices pioneered by the Internet businesses. Telcos are also acquiring IT technology services companies, building their own data centers, and offering their own cloud-based services.

2.7.1 Approaching the Traffic-Carrying Capacity of Fiber

The telcos are being challenged financially to invest in their networks as demand for capacity grows faster than their revenues (Fig. 2.7). What they need are technologies that help them add capacity to their networks at a lower cost-per-bit.

But telcos must address the problem of optical fiber reaching its capacity limit. They will no longer be able to keep scaling capacity as they have over the last 20 years. New technologies, perhaps even new fiber, will be needed to support the growth in bandwidth demand.

These problems will not be solved by continuing with the present setup; they require technology innovation—hence the rallying cry for solutions that extend optics beyond what is used today. This issue is

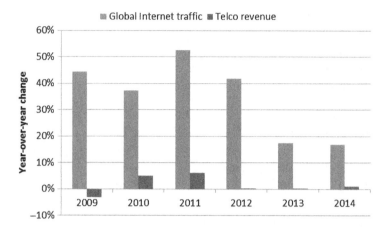

Figure 2.7 Telco revenues are not growing as fast as Internet traffic. Based on revenue figures from Ovum, traffic volumes from Cisco.

explored in Chapter 5, Metro and Long-Haul Network Growth Demands Exponential Progress.

2.7.2 Internet Businesses Have Their Own Challenges

Internet businesses' very success is forcing them to intervene to spur technological developments. What helps is that their growth and buzz means that their requests are met by a receptive vendor community.

Internet content providers have the financial clout to develop their own solutions or entice vendors to change direction and develop solutions for them ahead of the market. Indeed, the current period has brought about the most fundamental rethink of IT in decades—everything is up for review.

There are good reasons for this. One is that data center workloads are evolving quickly. The largest data center operators employ custom equipment configurations tailored to their core business tasks. Efficiency is key; the software algorithms and data center equipment the Internet content providers run are their engines of growth and they will do everything—including spurring the making of custom equipment—to achieve an edge.

The changing workloads are not just leading to a rethink regarding computing but also a reimagining of networking and data storage. And it is happening on many levels: where data centers are located, their interconnect requirements, how servers are designed, whether data transfers between the server processors and storage can be reduced to conserve power, and how best to connect servers, not just locally but across and between data centers.

2.7.3 Approaching the End of Moore's law

Evolving workloads are not their only cause for change. The fundamental driver of first the microelectronics industry and now the nanoelectronics industry, described by chip pioneer Gordon E. Moore in his famous law 50 years ago, is running out of steam. The chip giants are continuing to shrink the transistor, the elemental switch used to make integrated circuits. But the performance benefits described by Moore's law, as will be explained in Chapter 3, The Long March to a Silicon-Photonics Union, no longer scale.

Shrinking the transistor further is becoming exorbitantly expensive, which means the most basic economic engine of the information age is slowing down. The end of Moore's law is the backdrop to the changes taking place in the cloud, and inevitably it will make cloud optimization more challenging.

The Internet content providers are therefore thinking anew, and there is no agreement on the best approach to improving computing efficiency. It also explains why the web giants are developing industry initiatives such as Facebook's Open Compute Project [12], spurring the development of additional Ethernet rates, and creating industry consortia for such tasks as developing optics to connect chips [13]. In the future this period will be viewed as one of great upheaval, but it is also a period of significant opportunity, for new players and new technologies.

2.8 SILICON PHOTONICS: WHY THE TECHNOLOGY IS IMPORTANT FOR ALL THE LAYERS

Silicon photonics is a technology that has the potential to tackle applications across the layers: from long-distance, high-capacity telecom links at Layer 4 to compact optical devices doing battle with copper in the data center and down to Layer 1 distances.

Silicon photonics is coming to prominence because of its potential to tackle key emerging systems issues at the chip level (Layer 1) and for equipment (Layer 2 and Layer 3), namely their growing bandwidth requirements and power consumption.

The issues of connectivity and power consumption have become hugely important considerations regarding the data center. Interconnect costs in the data center and between data centers are considerable, and the market is already feeling the effects of Moore's law coming to an end. New technologies and approaches are needed to drive chip and system design forward—systems in the data center and equipment for Layer 4.

Silicon photonics, which shares a common base for optics and electronics, is one important technology in the toolkit of tricks engineers are considering for what is referred to as *Beyond* or *More than Moore's law*. This includes the promise of merging optics and electronics, thereby avoiding the challenges of bandwidth-reach and the power required to send signals between chips, and thus improving the overall system performance. Silicon photonics can also reduce total power consumption: long resistive electrical traces require energy to drive the signal, and the energy requirement goes up with data rate.

Yet another promise of silicon photonics is integration: the ability to combine numerous optical functions. Subsuming more and more functionality on a chip has been fundamental in the success of the chip industry. Optics is not like chip-making, as will be explained in Chapter 3, The Long March to a Silicon-Photonics Union, but optical integration has been fundamental in reducing system costs. And it is integration that will enable novel silicon photonics applications beyond just telecom and datacom, such as sensors, for example.

We see silicon photonics as a hugely promising optical technology that will play an integral role in the evolution of cloud computing and telecom at each of the four layers outlined. Silicon photonics is also a technology that benefits optical integrated designs and also the comingling of optics and electronics. Optical and electrical integration promises to lower cost—purchasing multiple parts is more expensive than buying fewer integrated devices because prices are marked up for each part, known as margin stacking.

In Chapter 3, The Long March to a Silicon-Photonics Union, we look at the origins of silicon photonics, how it differs from the chip industry, and how a slow but increasing union between electronics and optics is taking place.

Key Takeaways

- Telecom and datacom form a complex system that can be analyzed using a four-layered model.
- Each of the layers can be viewed as self-contained. The layers also share common attributes: all use communications, computing, and storage. What distinguishes the layers is distance or reach, pricing, and volumes.
- Optical technology plays an important role at each of the layers and, as such, so will silicon photonics.
- Silicon photonics will bring several important benefits to the layers including size, cost, and power consumption efficiencies for components and equipment.
- The greatest contribution of silicon photonics, however, will be to advance performance metrics beyond what is possible with existing technologies. With this will come new, more efficient system designs for networking, computing, and storage—attributes demanded by the telcos and especially the Internet content providers.

REFERENCES

[1] Gazzaniga MS. Tales from both sides of the brain: a life in neuroscience. New York: HarperCollins; 2015.

[2] Interview with the authors. Silicon photonics luminaries series, < http://www.gazettabyte. com/home/category/silicon-photonics-luminaries > .

[3] The connected vehicle—driving in the cloud. Gazettabyte, < http://www.gazettabyte.com/ home/2013/11/3/the-connected-vehicle-driving-in-the-cloud.html > ; November 3, 2013.

[4] Popular internet of things forecast of 50 billion devices by 2020 is outdated. IEEE Spectrum, < http://spectrum.ieee.org/tech-talk/telecom/internet/popular-internet-of-things-forecast-of-50-billion-devices-by-2020-is-outdated > ; August 18, 2016.

[5] Infinera targets the metro cloud. Gazettabyte, < http://www.gazettabyte.com/home/2014/11/6/ infinera-targets-the-metro-cloud.html > ; November 6, 2014.

[6] Farrington N, Andreyev A. Facebook's data center network architecture, < http://nathanfarrington.com/papers/facebook-oic13.pdf > .

[7] Cisco Global Cloud Index: forecast and methodology, 2014–2019, White Paper, < http:// www.cisco.com/c/en/us/solutions/collateral/service-provider/global-cloud-index-gci/ Cloud_Index_White_Paper.pdf > .

[8] The optical backplane is finally here. Will this change everything? LightCounting Market Research, Research Note; July 20, 2016.

[9] AT&T's 2014 annual report, < https://www.att.com/Investor/ATT_Annual/2014/downloads/ att_ar2014_annualreport.pdf > .

[10] Cisco's Visual Networking Index forecast: 2015−2020, < http://www.cisco.com/c/dam/ en/us/solutions/collateral/service-provider/visual-networking-index-vni/complete-white-paper-c11-481360.pdf > .

[11] Global threats to net neutrality. International New York Times, Saturday−Sunday April 11−12, 2015.

[12] Open Compute Project, < http://www.opencompute.org > .

[13] COBO acts to bring optics closer to the chip. Gazettabyte, < http://www.gazettabyte.com/ home/2015/4/30/cobo-acts-to-bring-optics-closer-to-the-chip.html > ; April 30, 2015.

CHAPTER 3

The Long March to a Silicon-Photonics Union

> You can't really understand what is going on now unless you understand what came before.
>
> *Steve Jobs [1]*

> By making things smaller, everything gets better. The performance of devices improves; the amount of power dissipated decreases; the reliability increases as we put more stuff on a single chip. It's a marvelous deal.
>
> *Gordon E. Moore [2]*

> Photons and electrons are like cats and dogs. Electrons are dogs: they behave, they stick by you, they are loyal, they do exactly as you tell them, whereas cats are their own animals and they do what they like. And that is what photons are like.
>
> *Mehdi Asghari [3]*

> The exciting thing about silicon photonics is that it is the only vector for scaling after [transistor] dimensional shrink.
>
> *Lionel Kimerling [4]*

3.1 MOORE'S LAW AND 50 YEARS OF THE CHIP INDUSTRY

It is rare for a trade magazine article to remain relevant for long, never mind 50 years after publication, nor is it often that an article's observation becomes a law—one whose consequences for the semiconductor industry the author foresaw decades in advance.

The article in question is by chip pioneer Gordon E. Moore and appeared in *Electronics* in 1965 [5]. Dr. Moore was the director of the R&D labs at Fairchild Semiconductor, an early maker of transistors. He also went on to cofound Intel and was the company's second CEO, following legendary chip pioneer Robert Noyce, the coinventor of the integrated circuit [1].

Silicon Photonics. DOI: http://dx.doi.org/10.1016/B978-0-12-802975-6.00003-X
Copyright © 2017 Daryl Inniss and Roy Rubenstein. Published by Elsevier Inc. All rights reserved.

Moore's article was written in the early days of integrated circuits. At the time, silicon wafers were 1 in. (25 mm) in diameter, and integrating 50 components on a chip was considered state of the art.

Moore observed that, in any period, there was an ideal number of components that could be included on a chip at a minimum cost. Add just a few more components and the balance would tip: the design would become overly complex, wafer yields would drop, and costs would rise.

His key insight, to become known as Moore's law, was that the complexity of an integrated circuit at this minimum cost was advancing over time. Moore expected this complexity to double each year for at least another decade.

He also predicted that, by 1970, the manufacturing cost per component would be a tenth of the cost in 1965. Extrapolating the trend further, Moore believed that by 1975, the number of components per integrated circuit for this minimum cost would be 65,000 components. Moore was overly optimistic, but only just: by 1975, Intel was developing a chip with 32,000 transistors.

Moore amended his law in 1975 to predict a doubling of complexity every 24 months. Moore's article talked about components, the basic elements of electrical circuits: transistors, resistors, and capacitors. But by 1975 the industry was focused on the transistor, which had become the building block for all the required circuit elements. The industry was also starting to alight on complementary metal-oxide semiconductor (CMOS) technology to make chips. And in the years that followed 1975, the period of complexity doubling became every 18 months.

Moore showed remarkable foresight regarding the importance of integrated circuits, especially when, in 1965, their merits were far from obvious. Such devices would bring a proliferation of electronics, he said, "pushing this science into many new areas."

He foresaw home computers (or at least "terminals connected to a central computer"), automatic control for automobiles, and even the mobile phone—personal portable communications equipment, as he called it.

The biggest potential of chips, he said, would be in the making of systems, with Moore highlighting computing, telephone communications, and switches. Fifty years on, such systems underpin datacom and telecom.

3.1.1 The Shrinking Transistor—But Not for Much Longer

The shrinking of the transistor has continued since Gordon E. Moore published his 1965 article, with remarkable technological and economic consequences [6].

The cost of making a transistor in 1965 was $30 in today's currency; in 2016 it is one billionth of a dollar. And in 2014, the semiconductor industry made 250 billion billion transistors, more than were made in all the years of the semiconductor industry up to 2011 [7].

As explained in the 50th anniversary issue on Moore's law in the *IEEE Spectrum* magazine, the miniaturization of transistors to date has been achieved with much engineering ingenuity and investment. Device yield—the proportion of working chips that come from a given wafer—has gone up from 20% in the 1970s to between 80% and 90% today. The size of the silicon wafers on which the chips are made has also increased, from 200 mm (8 in.) to 300 mm (12 in.). And while the lithography tools now cost 100 times more than they did 35 years ago, they also pattern the large wafers 100 times faster [7].

But the shrinking of the transistor cannot continue indefinitely, especially as certain transistor dimensions approach the atomic scale. Each new generation of CMOS process node is defined around a feature size of a transistor based on its gate length. Voltage is applied to the gate to turn the transistor off and on.

Leading chip companies are now using a 14-nm CMOS node to make their chips and are already targeting a 7-nm CMOS process [8]. The Belgium semiconductor research center, imec, is working on 7-, 5-, and 3-nm feature-size CMOS process technologies and claims that it sees a clear roadmap to achieve the 3-nm target [9].

But the industry is fast approaching the point where fewer benefits remain associated with smaller-sized transistors. Even the cost of making a transistor has stopped declining, with the transition point being around 28-nm CMOS (see Fig. 3.1) [10].

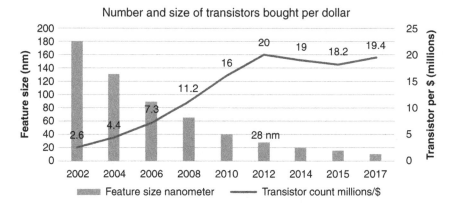

Figure 3.1 The changing economics of the chip industry. From The Linley Group.

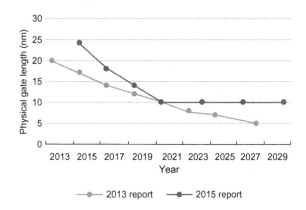

Figure 3.2 Moore's law coming to an end from 2021. Courtesy of ITRS.

Silicon manufacturing innovation will continue and transistors will shrink further. But no longer does the latest CMOS process deliver both improved performance and cheaper transistors.

The International Technology Roadmap for Semiconductors (ITRS) is an organization sponsored by the five leading chip manufacturing regions in the world: Europe, Japan, Korea, Taiwan, and the United States. In its 2015 report [11], it describes how the physical gate length of the transistor is set to remain at 10 nm from 2021 (see Fig. 3.2) ostensibly signaling the end of Moore's law. Smaller CMOS process nodes will continue to be developed, but the shrinking of the transistor gate length is almost done.

3.1.2 Post-Moore's Law: The Future of Optics and the Future of Chips

The industry may debate how many years of Moore's law are left, but it hardly matters. Moore's law has done a remarkable job, bringing the chip industry a half-century-long bounty such that it can now employ billions of transistors for integrated circuit designs [12].

To keep advancing computing performance beyond Moore's law, new thinking is required on many levels, from materials and circuit design to architectures and systems.

IBM illustrated this new thinking when it announced in 2014 a $3 billion research program over 5 years to extend chip development [13]. Areas it is exploring include quantum computing, neurosynaptic computing (a style of computing that mimics the brain) [14], III-V materials (combinations of column III and column V elements of the periodic table) to speed up electron mobility, carbon nanotubes, graphene, next-generation low-power transistors, and silicon photonics.

Like IBM, the industry group the ITRS recognizes that Moore's Law needs a rethink. The organization, as suggested by the name, regularly published future roadmaps for the chip industry based on the relentless scaling progress reflected in Moore's law. But in 2012 it became clear to the ITRS that the model was broken and a rethink was needed.

It resulted in the revamping of the organization, which became the ITRS 2.0, and it replaced its 17 International Technology Working Groups with seven focus teams (see Table 3.1) with an agenda of charting the chip industry's progress over the next 15 years: 2015–30 [11].

Only one of the seven focus teams is looking at techniques to extend Moore's law. The remaining six are concerned with a wide range of techniques and technologies that the industry will need to keep progressing computing and related technologies now that Moore's law is ending.

Interestingly, silicon photonics can play a role in six of the seven focus areas.

Mention of silicon photonics returns us to Gordon E. Moore's 1965 article. The article opened with a bold and prescient statement: "The future of integrated electronics is the future of electronics itself."

Table 3.1 The ITRS 2.0's Seven Focus Teams	
ITRS 2.0 Focus Teams	Comment
System Integration	Studies and recommends system architectures for the semiconductor industry for applications such as mobile, the data center and the Internet of Things
Heterogeneous Integration	The combination of differently manufactured components to create a packaged design with enhanced functionality
Heterogeneous Components	Micro and nano fabrication technologies that will be needed to create components for heterogeneous systems
Outside System Connectivity	Technologies to connect different parts of systems—such as wireless but also optical interconnect
More Moore	Developments to keep shrinking horizontal and vertical physical feature sizes for reduced cost and improved performance
Beyond CMOS	Identifying technologies to extend CMOS using heterogeneous integration, as well as new information processing approaches
Factory Integration	Tools needed to ensure that the semiconductor industry can continue to produce cost-effective designs in volume
Source: ITRS	

Can the same be said of photonics? Is the future of integrated photonics the future of photonics itself?

The answer is yes. While the bulk of optical components made today are not integrated [15], integrated devices will play an increasingly important role going forward. Integrated photonic circuits will be increasingly required as transmission speeds rise in the data center and for long-haul optical transport.

Photonic integrated circuits are different from electronic integrated circuits. Photonics uses several optical building blocks to make an integrated circuit (as described in Section 3.4), whereas electronics uses one, the transistor, as the basis of all chips. And while shrinking photonic functions does not deliver continual optical performance benefits as the shrinking of transistors has, integration of photonics using silicon promises to lower the cost of devices.

Refining the statement further: instead of integrated photonics, is the future of silicon photonics the future of photonics itself?

Here, as implied by IBM's investment program, the answer is that silicon photonics is bigger than photonics itself. Silicon photonics can be viewed as a semiconductor technology. A more accurate statement, therefore, is that the future of silicon photonics is the future of electronics itself, or at least one of its futures.

But there is another consequence, as we argue in Chapter 8, The Likely Course of Silicon Photonics, that the chip industry is the future of photonics itself.

3.2 HOW PHOTONICS CAN BENEFIT SEMICONDUCTORS

The semiconductor industry's billion-fold increase in transistor count over the past half-century has resulted in huge improvements to manufacturing techniques.

Large 300-mm silicon wafers—the size of a family pizza—are patterned and processed in controlled environments known as clean rooms to make chips with high yields. The industry also has advanced testing techniques to check that each chip is working while it is still on the wafer, as well as chip packaging technologies that can also benefit silicon photonics.

But now that Moore's law is faltering, chip manufacturing is becoming trickier. Transistors continue to shrink, but the speed at which chips are clocked is no longer rising.

Up till a decade ago, smaller transistors meant faster and cheaper transistors. Microprocessors—chips designed to run software—in particular made good use of the greater number of transistors to make more complex designs that could also be clocked faster. But this inevitably raised the chip's power consumption—an acceptable outcome for a costly device like a microprocessor until the thermal limits of integrated circuits started to be approached.

The industry then faced a tough decision: either to keep increasing the number of transistors or keep raising the clock speed at which the transistors are switched on and off, but not both. Chip designers chose to benefit from a greater transistor count with each process node and stopped ramping up the clock. As a result, processor clock speeds have stalled at a few gigahertz [16].

Processor designers are using the extra transistors to include several central processing units or processing cores on a chip to boost overall computing performance. Tasks are partitioned and distributed across the cores and processed in parallel. But some operations are inherently serial such that having multiple cores does nothing to accelerate the processing of code. Cores can thus be idle at times.

Designers must also contend with heat and power consumption issues as chips grow in complexity. And the continual need for higher processing performance raises system design issues. As chips advance, they need to be fed ever-increasing amounts of data, especially as the number of cores on a chip grows. EZchip, now part of Mellanox, is developing a networking chip that uses 100 ARM processor cores [17]. And a team at the University of California, Davis, Department of Electrical and Computer Engineering, has developed a chip containing 1000 independent programmable cores [18].

Multicore chip designs require faster internal data buses and more sophisticated hierarchical memory schemes, where the most important data is keep closest to a core in a high-speed on-chip store while the slower, larger memory located off-chip houses less urgent data (see Fig. 3.3). The chips also need to send and receive data, and this input−output is also getting faster over time.

All these requirements add to the chip's design complexity and overall power consumption. Fortunately, photonics can address many of these issues, in particular power and data input−output issues.

An optical waveguide, unlike a metal trace or a copper wire, does not suffer heating due to resistance. In turn, the resistance and capacitance of metal traces that connect different parts of a chip determine signal delay and hence signal speed, whereas the speed of light through an optical waveguide is a function of the speed of light divided by the waveguide's refractive index.

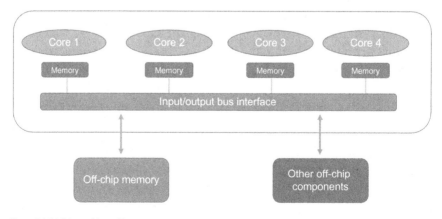

Figure 3.3 Multicore chip architecture.

And as the electrical signal speed increases, whether due to a faster input–output signal or a faster internal bus, the distance the signal travels diminishes. Optical transmission offers information-carrying capacity and distances that far exceed those of electrical transmission.

Indeed, the reach of optical communication offers the intriguing prospect of separating device functions to reduce the heat concentration without the equivalent of having to consume additional power to transmit electrons over distances in the electrical domain. In this way photonics can benefit chip design and enable novel system designs. An example of such a system is the disaggregated server, a topic discussed in Chapter 7, Data Center Architectures and Opportunities for Silicon Photonics.

For all these reasons, if integrated circuits are to evolve, comingling optical component technology with electronics makes sense.

3.2.1 Adapting Silicon for Optics

The optical component industry is much less mature than the chip industry and makes nowhere near the same unit volumes each year. And while there is a need for greater integration, the industry has no Moore's law driving relentless progress.

Optical components are made using materials such as indium phosphide, gallium arsenide, lithium niobate, and silica—a naturally occurring oxidized mineral source of silicon in the form of silicon dioxide. The idea of using silicon for photonics appeared radical at first, but the success of the semiconductor industry had long been noted by the optical industry, and silicon started to be researched for photonics and even as a single substrate to combine electronics with optics.

It turns out that silicon is an excellent material for optics. It is an abundant material and, thanks to the chip industry, is well understood. Silicon also has helpful material characteristics: it has good thermal properties as well as electrical ones. Silicon can be easily oxidized, as mentioned, and benefits from 300-mm silicon-on-insulator wafers developed for the CMOS industry. Such wafers are ideal for optics as they host silicon and silicon dioxide layers, ideal for optical waveguides to guide light as detailed in Section 3.4.1.

But to get silicon to work for optics has taken years of research effort. The good news is that silicon photonics is now moving to commercialization, and the design work that will deliver new devices and systems is in progress. These themes are discussed in Chapter 4, The Route to Market for Silicon Photonics.

3.3 SILICON PHOTONICS: FROM BUILDING BLOCKS TO SUPERCHIPS

The early research into optical components in the 1970s involved materials that now form the bedrock of optical communications.

Lithium niobate is one such material, used to modulate light for long-haul transmission since its optical properties—its refractive index—can be changed with the application of an electric field.

Modulation is used to transmit data over a transmission medium—fiber, optical waveguides, and even free space in the case of photonics—a task of key importance in communications. The modulator changes, or modulates, a particular parameter of a carrier signal. The carrier's signal role is to match the communication medium's characteristics. The modulator can change the carrier signal's amplitude, phase, or frequency, or combinations of these parameters. The modulator encodes the data to be sent by modifying one or more of the carrier signal's parameters. At the destination, the changes in the carrier signal are noted and the data is recovered. For optics, a modulator typically affects the amplitude or amplitude and phase of the light signal. Silicon photonics modulation is discussed in Section 3.4.2.

Other core optical materials include indium phosphide and gallium arsenide, III-V compounds that are ideal for generating, modulating, and detecting light. Indium phosphide has become the optical industry's monolithic photonic chip platform of choice due to its ability to implement all the main optical functions: laser, modulator, waveguide, and photodetector. Not surprisingly, it is the key-established technology that newcomer silicon photonics must compete with.

Research into the suitability of silicon for photonic circuits only began in the mid-1980s. What deterred progress until then was the fact that silicon does not lase; unlike indium phosphide and gallium arsenide, silicon is very inefficient at emitting light. Equally, silicon, unlike lithium niobate, does not exhibit the same linear electrooptic effect where its refractive index can be changed with an electric field. Thus silicon alone cannot be used as a laser source—a considerable negative—and at first glance, silicon looked unpromising as a modulator technology [19].

Once silicon photonics research started, a top-down concept—using silicon to make a monolithic superchip—was proposed. Such a chip, referred to as an optoelectronic integrated circuit, would combine optical functions with electrical circuits, as outlined by the founding father of silicon photonics, Professor Richard Soref [20]. This showed the potential of what could be offered: an integrated device combining electronics and optics (see Fig. 3.4).

The proposed superchip's optical functions were rich. The optical functions included waveguides, modulators, photodetectors, light sources, switches, and amplifiers. There were also built-in structures to aid the coupling of fiber to the chip.

The superchip's electrical circuits included a range of specialist transistor types. These included the HBT (Heterojunction Bipolar Transistor) and HEMT (High Electron Mobility Transistor), which are used for radio frequencies and millimeter wave frequencies, respectively. Another transistor technology is BiCMOS, which is used to process the received optical signals. BiCMOS combines two transistor

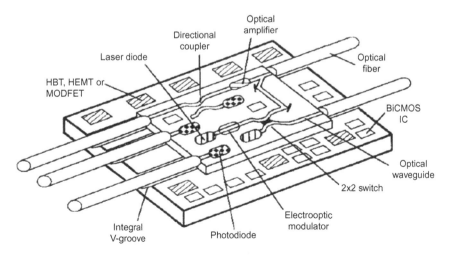

Figure 3.4 A concept optoelectronic integrated circuit superchip. © 2016 IEEE. Reprinted, with permission, from Soref RA. Silicon-based optoelectronics. In: Proceedings of the IEEE, vol. 81, no. 12; December 1993. p. 1687–1706. http://dx.doi.org/10.1109/5.248958.

types: bipolar and CMOS transistors. BiCMOS is a faster transistor technology than CMOS alone and was used to make advanced microprocessors for mainframes [16] and is used for radio designs.

But to be able to implement such a superchip, a bottoms-up approach would be needed: to develop the chip's basic building-block optical functions using silicon-based optics. First, techniques needed to be refined to improve the optical performance of each of the functions. And once the devices reached an acceptable performance, the challenge would be to combine them to create optical systems without compromising their individual optical performance.

These basic silicon photonics building blocks are discussed in the remainder of the chapter.

3.4 THE BUILDING BLOCKS OF SILICON PHOTONICS INTEGRATED CIRCUITS

There are two approaches used to combine optical functions to make integrated optical devices: monolithic integration and hybrid integration.

- *Monolithic integration* refers to a chip where the optical functions are all implemented using the same material system such as indium phosphide or gallium arsenide. Examples include an externally modulated laser that combines the laser and modulator or Infinera's latest 1.2 terabit photonic integrated circuit that combines hundreds of optical functions [21]. It is also possible to implement an optoelectronic IC monolithically. Such circuits typically combine the photonics with driver and receiver electronic circuitry.
- *Hybrid integration* involves constructing a photonic integrated circuit using two or more materials. The advantage of the hybrid approach is that the best material can be used for each individual optical function. But the different materials need to be combined, presenting its own design and manufacturing challenges.

Silicon photonics can implement hybrid or monolithic designs. Since silicon cannot lase, a silicon photonics monolithic chip is one that combines various optical functions but not the laser; if the circuit requires a laser, the design can only be a hybrid implementation. Silicon photonics has been used to make complex optical circuits and optical circuits combined with electronics. But such circuits typically require the laser and that means a hybrid design with the laser coupled to the chip.

However, using a hybrid technique known as heterogeneous integration, the laser can be bonded onto the silicon during the device's manufacturing. According to Professor John Bowers, bonding to silicon is attractive as it enables the integration of optical features that have not been widely integrated using any other platform [22]. These include not only lasers but also other active devices such as modulators and photodetectors, as well as passive functions such as optical isolators and circulators.

Moreover, using heterogeneous integration results in a single chip. The chip is made using hybrid integration in which two or more materials are used—e.g., silicon and indium phosphide—but the result is a chip with the laser integrated and aligned with other optical functions, just like a monolithic photonic integrated circuit.

In summary, silicon photonics chips can be implemented using monolithic or hybrid integration. However, if the photonic circuit requires a light source, the silicon photonics design is a hybrid one: either the laser is externally coupled, or bonded onto the silicon using heterogeneous integration. However, using heterogeneous integration results in a single die; it has all the attributes of a monolithic chip except two materials were used for its construction. Hence the overlap of heterogeneous integration with monolithic integration is shown in Fig. 3.5.

Having discussed how silicon photonics is used to make optically integrated circuits, it is time to discuss the building blocks used to make optical integrated circuits.

When sending data optically between two points, there needs to be a light source—a laser—a modulator, and a photodetector. In a

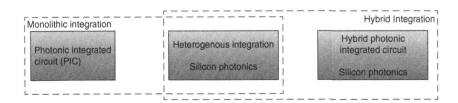

Figure 3.5 The different approaches to silicon photonics chip design.

photonic integrated circuit, optical waveguides are used to guide the light between the different optical functions. If data is to be sent in both directions, each end of the link has to have the transmitter (laser and modulator) and receiver (photodetector) functions. Such an optical system is referred to as a transceiver (see Appendix 1, Optical Communications Primer).

Of these optical building-block functions—the light source, waveguide, modulator, and photodetector—arguably the most important and simplest is the optical waveguide, as it forms the basis of the three other optical functions (see Fig. 3.6).

3.4.1 The Optical Waveguide

An optical waveguide routes light down a low-loss path. In effect the waveguide plays a similar role to optical fiber, except the waveguide is part of the silicon photonics chip. Like fiber, the light is confined using two materials with contrasting refractive indexes. The silicon-on-insulator wafers with their layers of silicon and silicon dioxide are ideal for constructing optical waveguides given the large refractive index contrast between the two: silicon has a refractive index of 3.5, while silicon dioxide's is 1.6 [23].

Two common waveguide structures are the channel or strip waveguide, and the rib waveguide.

The channel waveguide is a strip of silicon surrounded by oxide—cladding from above and a buried oxide from below. In contrast, the rib waveguide is shaped like an inverted T with a wide silicon base on which sits a narrower silicon ridge.

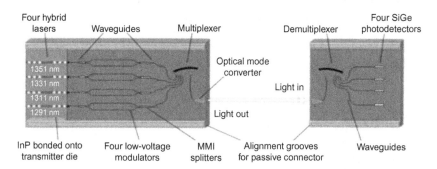

Figure 3.6 The optical building blocks used in a multiwavelength link. MMI, *multimode interference.* A. Alduino et al., "Demonstration of a high-speed 4-channel integrated silicon photonics WDM link with hybrid silicon lasers," Optical Society of America Integrated Photonics Research Silicon and Nanophotonics, Monterey, CA, paper PDIWI5 (July 25, 2010); http://bit.ly/e8zGqi.; courtesy of Intel.

The typical dimensions of the waveguide are sub-micron—several hundred nanometers wide and high. There is a trade-off between waveguide density on a chip and how lossy the waveguide is. Waveguides larger than a micron have lower loss, but their size limits the number that can be crammed on a chip. Submicron waveguides, in contrast, increase device density but are lossier.

The roughness of a waveguide's sidewalls is also a factor when it comes to light loss. The loss arises due to light scattering; smoothing the sidewalls reduces this loss. Another approach is crafting the geometry of the waveguide such that the strength of the light on the sidewalls is reduced. Typical losses for high-confinement waveguides are of the order of 2 dB/cm [24], although some companies report lower losses.

3.4.2 Modulation

Modulation is used to imprint information onto light prior to transmission, as described in Section 3.3. A silicon photonics modulator sits in front of the laser and either impedes or passes light. This is done by altering the modulator's optical properties—either its refractive index, which can be used to affect the speed of the light to enable interference between two light paths, or its absorption coefficient, a measure of the modulator's light blocking ability.

Desirable modulator characteristics include a small size to allow more optical channels to be crammed on a chip. The modulator also needs to be of low-power, operate over a wide range of wavelengths to enable wavelength-division multiplexing, have low insertion loss, and be of high-speed that typically means working at speeds of at least 50 Gb/s.

If it is the modulator's refractive index that is changed, two common modulator structures are used: the Mach–Zehnder interferometer and the ring resonator. These designs modify the refractive index and use phase changes in the light to create either constructive or destructive interference to induce light intensity changes.

Mach–Zehnder modulators have been optimized dramatically over the past decade and now work at high bit rates, but they are physically large—1 mm or more. Silicon photonics designs using a Mach–Zehnder structure have already been demonstrated for long-haul transmission.

Ring resonators are far more compact and have been shown to work at 50 Gb/s. But they are wavelength-specific and thermally dependent: a 1°C temperature change can detune the ring's resonance from the laser's wavelength. Ranovus is one silicon photonics player using ring resonator modulators for its optical transceiver products.

The second approach, changing the modulator's absorption coefficient, uses an electric field to vary the absorption coefficient of the material. Electroabsorption modulators use silicon germanium or indium phosphide and meet the small footprint requirement. Such modulators also have a small capacitance and achieve broadband operation.

Capacitance is important because it defines the modulator's maximum data rate as well as insertion loss: how much light power is lost passing through the modulator. Capacitance also dictates the modulator's extinction ratio: the ratio in decibels of the modulator's output light in the on and off states. The bigger the extinction ratio, the better. However, the resulting capacitance generally requires a trade-off between the modulator's various performance metrics [25].

The characteristics of the main three silicon photonics modulator types are summarized in Table 3.2.

Table 3.2 Optical Modulator Types			
	Mach–Zehnder	Ring Resonator	Electroabsorption
Optical mechanism	Refractive index	Refractive index	Absorption coefficient
Comments	Speeds of 50 Gb/s demonstrated but has relatively large dimensions—1 mm—that limit future scaling	Small size and reduced capacitance compared to the Mach–Zehnder. Also demonstrated at 50 Gb/s. Its lower capacitance means a reduced dynamic power consumption	Small footprint and small capacitance
	The modulator also has a relatively large capacitance, which limits its speed of operation and future power consumption reductions	The design must be thermally stable to ensure that its resonance frequency does not drift	Demonstrated operating above 50 Gb/s
			Has a level of thermal robustness and works over a broadband of wavelength (tens of nanometers). But the design drifts with temperature, ~ 0.8 nm/°C

3.4.3 Photodetection

Photodetectors are a critical component at the optical receiver. This optical functional block converts the received light into an electrical current. Compactness, as with the modulator, is important, as is its bandwidth or speed of operation. A key defining performance parameter of the photodetector is its responsivity—the electric current output that results as a function of a unit of optical power (amp/watt) on the photodetector.

To add the photodetector function to silicon, germanium is added through doping the chip. Alternatively, an external material such as an indium phosphide photodetector can be added using hybrid or heterogeneous integration.

3.4.4 The Light Source

The light source is the critical component of any optical communication link and it is the silicon's biggest shortfall. Because silicon is an indirect bandgap material, much work has been undertaken to enable silicon to lase. For now, silicon photonics players have circumvented the material's limitations and use either an external laser coupled to the silicon photonics chip or use hybrid integration by bonding a III-V laser.

One advantage of using an external laser is its separation from the chip: a laser generates heat, and separating it makes temperature management easier. Lasers are also a mature optical component technology with a wide choice of suppliers.

Adding a laser to the photonic chip represents a manufacturing challenge; coupling a laser to the chip has ramifications regarding a design's optical performance and cost. For an optical transceiver design, the laser-attach approaches include coupling a discrete laser to the chip (attached on-chip or off-chip via fiber) or adding a lasing material on the wafer using hybrid integration (see Fig. 3.7).

A standalone laser is independent of the silicon chip fabrication process. Because the performance and reliability of the discrete laser are already known, the design only needs to be tested once the laser is attached. However, accurate alignment of the laser to the chip is a must, and this is cumbersome, especially as the number of channels (laser count) grows.

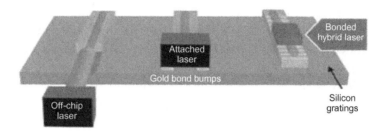

Figure 3.7 Schematic illustrating standalone and hybrid lasers used with silicon photonics chips. From Photonics Spectra (Intel).

The second, hybrid approach, also known as heterogeneous integration, involves attaching indium phosphide to the silicon wafer. Either indium phosphide wafers 50 mm and 75 mm in diameter are bonded to the 300-mm silicon wafer, or "chiplets"—slivers of indium phosphide—are located where needed on the wafer before the two materials are processed in situ [25,26].

The hybrid approach uses wafer-scale processing, with alignment of the laser to the waveguide part of the wafer fabrication process. This makes hybrid better suited for larger-volume applications and for circuits that have higher channel counts. Note that it also leads to lower coupling loss (and, therefore, lower power consumption for the circuit), lower cost, and the opportunity to leverage the silicon structure for more complicated laser functionality such as wide tunability. And since the lasers are buried, they are inherently hermetic. However, the laser cannot be tested until the entire device is assembled.

A third approach, promising the highest efficiency and most suited for volume applications, is to grow indium phosphide directly onto the silicon wafer. Silicon and indium phosphide have different crystal structures such that growing one on the other gives rise to defects. However, researchers are working to localize the defects to create working lasers. Such a monolithic approach remains a challenge, and further research is needed before this is ready for commercialization [25,27].

The light source's impact on overall cost and optical performance is summarized in Table 3.3.

Table 3.3 Impact on Cost and Performance Using a Discrete, Hybrid, or Monolithic Laser Approach

	Bill of Materials Cost	Processing Cost	Testing Cost	Assembly Cost	Performance Issues
Standalone laser	Purchased device, may need hermetic package, a lens, and isolator	Two separate processes	Test laser, test device, and test laser attached to device	Couple laser to silicon waveguide	Known, trusted, understood, can put laser anywhere, and thermally manage
Hybrid laser	Material cost, intrinsically hermetic, lens and isolators not needed	Laser wafer material process and waveguide processes	Test device at wafer level	Wafer bonding or chiplet placement	Performance only known on device completion. Thermal management
				Laser integrated with waveguide in wafer processing	
Monolithic laser	Material cost. Indium phosphide growth	One process	Test device at wafer level	Laser integrated with waveguide in wafer processing	Performance only known on device completion. Thermal management

Devices commercialized by Acacia, Cisco, and Luxtera all use discrete lasers. This is currently the most pragmatic approach and has enabled the companies to get products to market.

The hybrid approach promises lower bill-of-material costs, wafer-scale testing, and fewer assembly steps overall, which reduces overall cost. But the processing is a challenge. The laser is part of the chip and subject to the chip's high temperature, raising thermal management issues. And yields will be low initially, harming the overall economics. Juniper Networks, Skorpios, and Intel are all pursuing the hybrid approach. Intel's first commercial products using heterogeneous integration are 100 Gb optical transceivers [28]. As Intel points out, no alignment is needed; the laser is aligned during the silicon chip's manufacturing using photolithography.

3.4.5 Fibering the Chip

The photonics chip has to be connected optically to the outside world, independent of how the laser is attached. Fiber is used to bring light to and from the chip. In addition, fiber is also an option to attach a laser to the silicon chip.

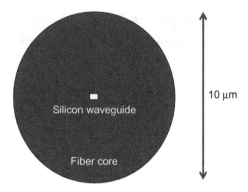

Figure 3.8 Single-mode optical fiber core diameter compared to the tiny silicon waveguide.

Attaching fiber to the chip remains the biggest manufacturing challenge for silicon photonics, because optical loss must be minimized in the coupling process to ensure acceptable optical performance.

Coupling fiber to a silicon waveguide is challenging because of the different densities of the confined light, referred to as the mode field diameter mismatch. Light is tightly confined in silicon due to the large refractive index difference between the silicon core and the silica cladding. The resulting core is less than 1 µm in diameter. In contrast, the optical fiber core is about 9 µm with a considerably smaller refractive index difference between the germanium-doped core and the silica cladding. Fig. 3.8 shows the core size differences and the challenge.

Two approaches are used to get light on or off the chip: edge coupling and grating coupling.

Edge coupling is used to butt-couple fiber—or a laser for that matter—to the silicon waveguide. Here the light from the fiber propagates in the same plane as the waveguide. Given the large refractive index difference between the chip's silicon core and the silica cladding, the beam of light—its mode field—is considerably smaller in a silicon chip than in the optical fiber. A mode-matching technique is needed, such as the inverse taper shown in Fig. 3.9. Here, the silicon waveguide is tapered to gradually match the transition between the two mode field diameters.

As the name implies, a grating coupler requires a grating to be created within the silicon waveguide (Fig. 3.10). The gratings can be broadband, and the approach is attractive because it can be designed

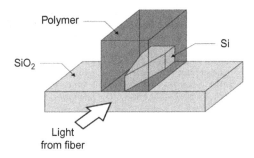

Figure 3.9 An inverse taper structure used to expand the silicon waveguide mode field diameter to match that of the optical fiber. McNab SJ, Moll N,Vlasov YA. Ultra-low loss photonic integrated circuit with membrane-type photonic crystal waveguides. Opt Express 2003;11:2927–39.

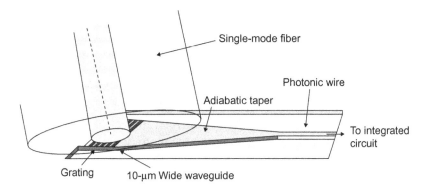

Figure 3.10 The grating coupling approach. Journal of Lightwave Technology [29]. Van Laere F, Bogaerts W, Taillaert D, Dumon P, Van Thourhout D, Baets R. Compact focusing grating couplers between optical fibers and silicon-on-insulator photonic wire waveguides. In: Optical fiber communication conference and exposition and the national fiber optic engineers conference, OSA technical digest series (CD), Optical Society of America Paper OWG1; 2007.

with little back-reflection. The coupler can be placed anywhere on the chip, the spot is large and hence it enables easy alignment with the fiber, and the grating supports wafer testing. Here, light is delivered to the grating vertically, perpendicular to the waveguide plane.

Neither coupling technique has yet demonstrated a proven optical performance advantage to become the de facto approach. Designers thus continue to assess which approach best suits their particular designs.

In summary, Table 3.4 lists the main optical functions and their performance issues.

Having introduced the key silicon photonics building blocks, the companies and individuals that have helped bring these optical functions

Table 3.4 Optical Functions and Their Performance Issues		
Optical Function	CMOS-Compatible Processing	Optical Performance
Laser	Cannot be implemented in silicon. Requires a III-V compound such as indium phosphide either bonded to the silicon wafer or an off-chip laser coupled to the silicon chip	1310 and 1550 nm wavelengths demonstrated. Multichannel design—wavelength-division multiplexing—has also been demonstrated
Modulator	Can be implemented in silicon. Also other materials such as indium phosphide can be bonded and used	Low power, 50 gigabit per second modulation speeds demonstrated. Different modulators available, each with its own merits and shortfalls
Waveguide	The size of the silicon waveguide and hence the interface density (how many can be placed side by side) is determined by the diffraction limit of light	Waveguide surface roughness can impact signal loss. There is also insertion loss when coupled to fiber due to mode field mismatch
Photodetector	Requires germanium doping	No obvious issues

to maturity are discussed in Chapter 4, The Route to Market for Silicon Photonics. The history is a recent one, highlighting how rapid silicon photonics' progress has been. The commercial impact of silicon photonics to date and the challenges facing the technology are also discussed in Chapter 4, The Route to Market for Silicon Photonics.

Key Takeaways

- Shrinking the transistor no longer gives the returns that the semiconductor industry has benefited from for decades. Moore's law is approaching its end.
- In the next era, known as More than Moore's law, integration of multiple technologies will become prevalent. Photonics will be one such technology, arguably a crucial one.
- Silicon photonics offers the most elegant approach to combining logic and photonics.
- Silicon's key shortfall is that it cannot generate light. This requires external coupling of laser chips or the bonding of III-V materials onto the silicon wafer. Research work continues to develop ways to enable silicon to lase, but that is some way off from being a commercially viable option.
- All the other main optical functions—waveguides, switching, modulation, and photodetection—can be implemented using a CMOS-compatible process.
- The development of silicon photonics from the research lab to the stage where it will be a commercial technology has been a considerable undertaking.

REFERENCES

[1] Berlin L. The man behind the microchip: Robert Noyce and the invention of silicon valley. Oxford: Oxford University Press; 2005.

[2] The law that's not a law, Interview with Gordon Moore. IEEE Spectrum, < http://spectrum.ieee.org/computing/hardware/gordon-moore-the-man-whose-name-means-progress >; April 2015.

[3] Silicon Photonics. A Q&A with Kotura's CTO. Gazettabyte, < http://www.gazettabyte.com/home/2012/10/16/silicon-photonics-qa-with-koturas-cto.html >; October 16, 2012.

[4] Interview with the authors. Silicon photonics luminaries interview. Gazettabyte, < http://www.gazettabyte.com/home/category/silicon-photonics-luminaries >.

[5] Moore GE. Cramming more components onto integrated circuits. Electronics 1965;38(8).

[6] Manners D, Makimoto T. Living with the chip. London: Chapman & Hall; 1995.

[7] The multiple lives of Moore's law, Chris Mack. IEEE Spectrum, < http://spectrum.ieee.org/semiconductors/processors/the-multiple-lives-of-moores-law >; April 2015.

[8] IBM discloses working version of a much higher-capacity chip. New York Times, < http://www.nytimes.com/2015/07/09/technology/ibm-announces-computer-chips-more-powerful-than-any-in-existence.html?_r = 0 >; July 9, 2015.

[9] Imec gears up for the Internet of Things economy. Gazettabyte, < http://www.gazettabyte.com/home/2016/3/3/imec-gears-up-for-the-internet-of-things-economy.html >; March 3, 2016.

[10] The end of Moore's law. The Economist, < http://www.economist.com/blogs/economist-explains/2015/04/economist-explains-17 >; April 19, 2015.

[11] International Technology Roadmap for Semiconductors 2.0, 2015 Edition, Executive Report, < http://www.semiconductors.org/main/2015_international_technology_roadmap_for_semiconductors_itrs/ >; July 2016.

[12] Altera's 30 billion transistor FPGA. Gazettabyte, < http://www.gazettabyte.com/home/2015/6/28/alteras-30-billion-transistor-fpga.html >; June 28, 2015.

[13] IBM announces $3 billion research initiative to tackle chip grand challenges for cloud and big data systems, < https://www-03.ibm.com/press/us/en/pressrelease/44357.wss >; July 10, 2014.

[14] Introducing a brain-inspired computer, TrueNorth's neurons to revolutionize system architecture. IBM Research, < http://www.research.ibm.com/articles/brain-chip.shtml >.

[15] Is silicon photonics a disruptive technology? Market opportunity for optical integration technologies, LightCounting Market Research report; January 2016.

[16] The next era of enterprise processor design. IBM's Charles Webb, keynote talk. In: The 45th international symposium on microarchitecture, Micro 2012. < https://www.youtube.com/watch?v=emlZ43n9zP0 >.

[17] EZchip packs 100 ARM cores into one networking chip. Gazettabyte, < http://www.gazettabyte.com/home/2015/2/23/ezchip-packs-100-arm-cores-into-one-networking-chip.html >; February 23, 2015.

[18] World's first 1,000-processor chip. UC Davis, < https://www.ucdavis.edu/news/worlds-first-1000-processor-chip >; June 17, 2016.

[19] Reed GT, Headley WR, Jason Png CE. Silicon photonics—the early years. In: Proceedings of SPIE, vol. 5730; March 7, 2005.

[20] Soref R. Silicon-based optoelectronics. Proceedings of the IEEE 1993;81(12):1687−706.

[21] Infinera goes multi-terabit with its latest photonic IC, < http://www.gazettabyte.com/home/2016/3/30/infinera-goes-multi-terabit-with-its-latest-photonic-ic.html >; March 30, 2016.

[22] Heterogeneous integration comes of age. Silicon photonics luminaries series, John Bowers. Gazettabyte, < http://www.gazettabyte.com/home/2016/8/28/heterogeneous-integration-comes-of-age.html >; August 28, 2016.

[23] Fundamentals of silicon photonic devices. Kotura, < http://www.mellanox.com/related-docs/whitepapers/KOTURA_Fundamentals_of_Silicon_Photonic_Devices.pdf >.

[24] Chrostowski L, Hochberg M. Silicon photonics design: from devices to systems. Cambridge: Cambridge University Press; 2015.

[25] Vivien L, Pavesi L, editors. Handbook of silicon photonics. Boca Raton, FL; New York, London: CRC Press; 2013.

[26] Heck MJR, Bowers JE. Hybrid and heterogenous photonic integration. Handbook of silicon photonics. Boca Raton, FL: CRC Press; 2013 [chapter 11].

[27] Imec and Ghent university demonstrate first laser arrays monolithically grown on 300 mm silicon wafers, < http://www2.imec.be/be_en/press/imec-news/ghent-university-nature-si-photonics-indium-phosphide-lasers.html >.

[28] Intel's 100 gigabit silicon photonics move. Gazettabyte, < http://www.gazettabyte.com/home/2016/8/21/intels-100-gigabit-silicon-photonics-move.html >; August 21, 2016.

[29] McNab SJ, Moll N, Vlasov Y. Ultra-low loss photonic integrated circuit with membrane-type photonic crystal waveguides. Optics Express 2003;11(22):2927–39.

CHAPTER 4

The Route to Market for Silicon Photonics

A technology does something. It executes a purpose.

W. Brian Arthur [1]

Once you can do something in silicon and do it adequately well, it tends to displace everything else from the majority of the market. We have seen that over and over again, and I don't think it will be any different in the optics world.

Michael Hochberg [2]

4.1 THE TECHNOLOGY ADOPTION CURVE

How does a new technology arise and become adopted in the marketplace?

In his book on the nature of technology, W. Brian Arthur talks about how new technology builds on existing ones, identifying a process he calls *internal replacement* [1].

Once a technology is adopted, its performance is pushed to deliver more. If it is not operating close to its limit, it is, by definition, being used inefficiently. When competition is severe—as it is for the optical component and equipment vendors addressing the datacom and telecom markets—even a small edge can pay off handsomely, says Arthur.

But you can only push a technology so far before part of the system hits a barrier. One way to overcome this limitation is to replace the impeding component—a subtechnology—with a better one. This improved component will likely require adjustments in other parts of the system to accommodate it. Arthur cites how moving from a wooden to a metal aircraft frame in the 1920s and 1930s led to a rethink of aircraft design.

Silicon Photonics. DOI: http://dx.doi.org/10.1016/B978-0-12-802975-6.00004-1
Copyright © 2017 Daryl Inniss and Roy Rubenstein. Published by Elsevier Inc. All rights reserved.

Internal replacement improves a system's parts and subparts and impacts all levels of its hierarchical design.

Similarly, if silicon photonics is to be adopted widely, it must internally replace existing component technologies, delivering an edge that established technologies cannot match. Its impact could be as a component, or for more advanced parts of a system. Silicon photonics could even lead to new, novel systems given its compactness, interconnect reach, and integration potential.

In turn, technologies that are adopted in the marketplace go through a recognizable development cycle, and silicon photonics is proving no different. Market research and advisory firm Gartner has identified a four-stage technology development cycle, shown in Fig. 4.1.

The four-stage cycle starts with an *innovation*: a development or an advancement that builds on existing knowledge to create the new technology [3]. Research embraces and builds on this innovation, sparking enthusiasm and a market expectation once the technology's potential is recognized. At this next stage, the market explores potential applications, and start-ups may enter the market with an innovative product using the technology. The enthusiasm feeds on itself and soon expectations become inflated. But eventually this crescendo of interest reaches a *peak* and is followed, inevitably, by deflation.

Several reasons account for this *disillusion* phase. Developing a technology sufficiently to enter the marketplace is challenging; setbacks

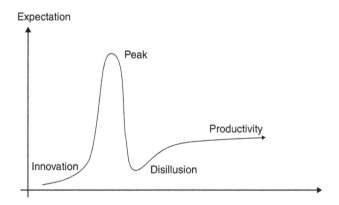

Figure 4.1 Gartner's four-stage technology development cycle. Based on Gartner.

and delays are inevitable. Equally, the marketplace continues to evolve such that the technology's initial promise may lose some luster. But like the peak of expectation, this disillusionment period also passes; successful technologies are adopted after all.

What then follows is the long hard slog to bring the technology to its *productivity* phase, and this is the current status of silicon photonics.

This chapter details the key events that have led silicon photonics to reach the productivity phase. Four notable early silicon photonics product case studies are detailed. The chapter also addresses the challenges facing silicon photonics and the current state of the industry.

4.2 A BRIEF HISTORY OF SILICON PHOTONICS

Silicon photonics may first have been thought about in the mid-1980s, but the rapid strides made in its development have occurred since 2000. Fig. 4.2 highlights key silicon photonics developments during this period.

In the following sections we discuss the events in Fig. 4.2, from earliest to latest.

NTT [4] and Cornell University [5] developed an inverted taper fiber coupler design in 2002 and 2003, respectively. The taper is used to match light between the fiber core and the much smaller silicon photonics waveguide, as discussed in Chapter 3, The Long March to a Silicon Photonics Union.

Intel announced its first silicon photonics breakthrough, a 1-GHz modulator, in 2004. The world's leading chipmaker decided that silicon photonics would be a core technology, and the company's R&D engineers, over several years, pushed laser, modulator, and photodetector development for the technology.

Mario Paniccia, who headed Intel's silicon photonics efforts at the time, spotlights a particularly creative period between 2002 and 2008. During that time, his Intel team had six silicon photonics papers published in science journals *Nature* and *Nature Photonics*, and it held several world records. These were for the fastest modulator, first at 1 Gb/s, then 10 and finally at 40 Gb/s; and the first pulsed and continuous-wave

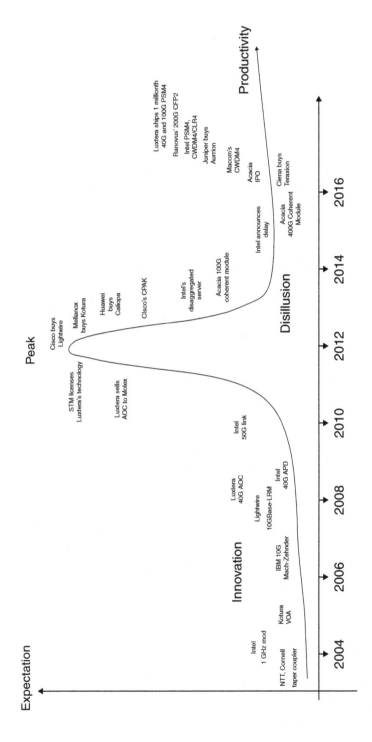

Figure 4.2 Silicon photonics' technology development cycle.

Raman silicon laser. Additional records followed: the first hybrid silicon laser based on work with the University of California, Santa Barbara, and the fastest silicon germanium photodetector operating at 40 Gb/s. These piece parts were required before Intel moved on to start designing optical transceivers [6]. These achievements were all in one place, in labs within 100 yards of each other, says Paniccia: "You had to pinch yourself sometimes." [7]

Intel was not alone. IT giants IBM, HP, and Sun Microsystems (since acquired by Oracle) used their research and development labs to explore the technology also. IBM developed a 10-Gb/s Mach–Zehnder silicon photonics modulator in 2007 and the first 40-Gb/s germanium receiver in 2010. But Intel was the main cheerleader and did most to highlight the technology's potential, heightening industry expectation toward its peak.

The innovation era also spawned three significant silicon photonics start-ups: Kotura, Lightwire, and Luxtera. Kotura was first to market with its silicon photonics–based variable optical attenuator product in 2005. Lightwire launched a 10-Gb Ethernet optical module in 2008, and Luxtera followed in 2008 with a 40-Gb/s active optical cable (AOC) for linking equipment within the data center.

Two of these three early start-ups were also notable for their high market valuations, given that there were only a few silicon photonics products and little revenue at the time. Cisco Systems paid $271 million for Lightwire in 2012, making it the most expensive silicon photonics start-up acquisition to date [8]. And in 2013 datacom equipment maker Mellanox Technology acquired Kotura for $82 million [9]. Meanwhile, Luxtera sold its AOC business to Molex in 2011 [10].

In hindsight, the first wave of start-up acquisitions marked the peak of enthusiasm for silicon photonics on the technology adoption curve. Innovation and industry developments have continued, but the spate of acquisitions, and the sums paid, put the spotlight firmly on silicon photonics during this period.

Financial analysts also alighted on the technology's potential, warning leading optical component companies such as Finisar that not having a silicon photonics strategy would harm their businesses [11].

Leading Chinese equipment maker Huawei bought a small Belgium silicon photonics start-up Caliopa for an undisclosed fee in 2013 [12], although reports suggested it was $5 million, while in 2014 Macom acquired BinOptics for $230 million [13]. The amount paid was not solely for the company's silicon photonics since BinOptics also makes lasers, a valuable technology asset. Macom also bought Photonic Controls, a silicon photonics design company [14].

Other notable product developments included Intel announcing its "disaggregated" server with silicon photonics earmarked to play an integral role in its design [15]. Recall that the server is the computing platform used in huge volumes in large-scale data centers, while disaggregation refers to separating the processing, memory, and storage functions as part of an industry reassessment as to how best to architect servers.

Cisco also launched its custom CPAK 100-Gb/s transceiver, its first silicon photonics product based on Lightwire's technology [16].

Start-up Acacia Communications began selling a silicon photonics–based 100-Gb/s, low-power, pluggable coherent transceiver for metro and long-haul applications [17]. A year later, Acacia announced a follow-on coherent module that supports data rates of 200, 300, and 400 Gb/s [18].

The burst of acquisitions of 2012 and 2013 was followed by a quieter phase, dampening industry expectations. Adding to the sense of lost momentum was the announcement in early 2015 by Intel—which has done most to trumpet the technology—that its silicon photonics–based product plans would slip a year [19].

IBM, like Intel, has been developing silicon photonics technology for over a decade. It announced in 2015 a nearly complete 100-Gb/s transceiver using wavelength-division multiplexing technology. However, while the company said it would use the transceiver for its computing systems, no schedule was given. IBM added that its design would need wider adoption to become cost-competitive with existing optical component technologies [20].

Accordingly, 2014 through the first half of 2015 was silicon photonics' era of disillusionment. Now, the industry has exited its quiet period. Ciena bought Teraxion's silicon photonics team for $32 million in early 2016 [21], while Juniper Networks acquired start-up Aurrion for $165 million in August [22]. In between, Acacia Communications achieved a successful initial public offering, raising $105 million for the company [23].

More and more companies are adopting the technology and announcing first products. For inside the data center, Macom announced a 100-Gb/s CWDM4 (coarse wavelength-division multiplexing 4) module [24] as has the firm Kaiam [25], while Lumentum detailed its PSM4 (parallel single mode 4) module [26]. All three designs use silicon photonics in a pluggable QSFP28 form-factor module. Start-up Ranovus detailed a 200-Gb/s CFP2 direct detection optical module with a reach of up to 130 km for Layer 4 data center interconnect applications [27]. Intel too announced its first silicon photonics optical modules for PSM4 and CWDM4/CLR4 [28].

All these companies are addressing the challenges that need to be overcome to make silicon photonics a cost-competitive alternative to the established optical technologies. Companies and start-ups will fail even as the overall silicon photonics proposition continues to take root, but that is the case with any technology transition.

4.3 FOUR COMMERCIAL SILICON PHOTONICS PRODUCT CASE STUDIES

Several silicon photonics products are being sold in volume, including one sold since 2005. Studying these products helps identify silicon photonics' merits as well as the challenges companies face when bringing the technology to market.

4.3.1 Kotura's Variable Optical Attenuator

Kotura's variable optical attenuator was the first silicon photonics—based product to be shipped in volume. A variable optical attenuator is used to trim a fiber's optical signal power levels. Applications include leveling the power exiting an optical amplifier across a fiber's spectrum, and protecting a photodetector at a receiver from being overwhelmed by too strong a signal (Fig. 4.3).

Kotura introduced its variable optical attenuator in 2005. One decade later, 10000 units were being shipped each week. Kotura was acquired by Mellanox Technologies in 2013.

Figure 4.3 Mellanox's (Kotura's) variable optical attenuator. Courtesy of Mellanox Technologies.

The product's success stems from its responsiveness. Its signal leveling and protection occur in under a microsecond, far faster than equivalent variable optical attenuators based on other technologies. The product is thus ideal to control transient—spurious—light. Also Mellanox's variable optical attenuator is highly reliable as there are no moving parts, unlike other approaches. And given the volumes made, its low cost and price make it a formidable competitor.

This is silicon photonics' first example of a product trumping the competition by having superior optical performance. For Kotura, it was also a shrewd move. The company made a simple yet performance-leading product for telecom while gaining valuable production experience as well as revenues, a winning combination for a start-up.

4.3.2 Luxtera's 40-Gb Active Optical Cable (AOC)

Luxtera launched its 40-Gb/s AOC using silicon photonic—based transceivers in 2008. AOCs are used to connect equipment over short reaches: from 10 m to a few hundred meters, although the bulk of links are 30 m or less.

Standard form-factors are used at each end of an AOC slot into the equipment's faceplate, but the transceiver optics are embedded in housing that is part of the cable. Since vendors control each end of the link, they can use proprietary—nonstandardized—optical designs.

Luxtera's 40-Gb product used a single-mode fiber ribbon cable: four fibers to transmit and four to receive. For the transmitter, Luxtera cleverly used a single laser shared across the four channels, saving costs and reducing power consumption. Each of the four channels has its own silicon photonics modulator. The resulting design, claimed Luxtera, offered the lowest-power option for data center interconnects, consuming 20 mW/Gb, 30% less than competing approaches [29].

Luxtera spent several years exploring applications where it could use silicon photonics to produce superior products. "We ship product, and to ship you need to have product differentiation against a pretty competitive landscape," explains Luxtera executive Brian Welch [30].

The start-up studied short-reach links but was put off by the strong competition from vertical-cavity surface-emitting lasers that operate at 850 nm. Luxtera also examined wavelength-division multiplexing but concluded that while silicon photonics faced no technical hurdles in addressing this application, it offered no advantage compared to indium phosphide–based products.

This led Luxtera to alight on a parallel single-mode fiber design for the data center—in effect a 40-Gb PSM4 design before the PSM4 multisource agreement was conceived (see Appendix 1: Optical Communications Primer).

It is important to stress that the AOC's optical performance is not specific to silicon photonics; a similar design could be made using indium phosphide. But Luxtera used its silicon photonics expertise and achieved first-mover advantage.

Luxtera has since developed a 56-Gb/s four-channel PSM4 cable, having discontinued its 40-Gb AOC. Luxtera sold its AOC business to Molex in 2011 [10].

The company also makes a 100-Gb/s (4 lanes, each at 25 Gb/s) PSM4 transceiver as well as selling the internal chips, dubbed optical engines (Fig. 4.4). In September 2016 it announced that it had shipped its one millionth PSM4 transceiver—the sum of all its 40- and 100-Gb/s shipments, including AOCs and standalone optical transceivers [31].

4.3.3 Cisco Systems' CPAK

Cisco Systems acquired Lightwire, the silicon photonics start-up, in 2013 after realizing its next-generation switching and routing

Figure 4.4 Luxtera's 100-Gb PSM4 silicon photonics chip. From Luxtera.

equipment would be delayed waiting for the industry to start selling 100-Gb pluggable modules in the CFP2 form factor.

Having a 100-Gb CFP2-sized module was important for Cisco to provide the necessary input−output traffic to feed its 400-Gb network processor. A network processor is a specialized chip that processes and classifies Internet Protocol packets at line rates up to 400 Gb/s. Cisco designs its own network processor ASICs (application-specific integrated circuits) to gain a competitive edge. We will meet network processors again in Chapter 7, Data Center Architectures and Opportunities for Silicon Photonics.

Because the industry was late with the 100-Gb CFP2 optical module specification, Cisco decided to buy Lightwire to make its own 100-Gb custom pluggable module, dubbed the CPAK (Fig. 4.5). This sounds like a risky strategy because making its own module would also take time, but Cisco had already been working closely with Lightwire before the acquisition.

"By the time someone could provide us with the CFP2 optics that could achieve the input−output density, that network processor silicon investment would have gone stale," says Russ Esmacher, a Cisco Systems executive [30].

The CPAK transceiver is not solely silicon photonics−based. The four channels, each operating at 25 Gb/s, use indium phosphide lasers

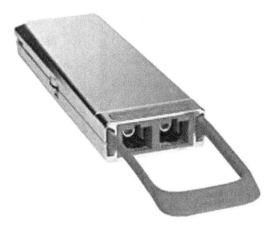

Figure 4.5 Cisco's CPAK. From Cisco.

and photodetectors, but the modulators and optical waveguides are based on silicon photonics.

By making the CPAK, Cisco quadrupled the packet-processing performance of its line card while keeping power consumption fixed. This is a common equipment requirement: doubling or quadrupling the performance of a card and system without increasing power consumption, an expectation set by Moore's law and largely fulfilled.

The equipment maker upgraded its 100-Gb line card to one with a 400-Gb network processor and four CPAKs, each implementing the IEEE 100GBase-LR4 100-Gb Ethernet standard that has a 10 km reach.

Cisco gained an enabling technology by buying Lightwire, resulting in a year's lead on its switch-vendor competition. Such platforms sell for far more than the cost of the individual modules. Cisco also gained valuable experience with an optical technology that it now has in-house, a photonic complement to its ASIC expertise.

4.3.4 Acacia Communications' Coherent Transceivers
Acacia Communications is one company targeting the challenging long-distance optical transmission market.

Founded in 2009, Acacia has made the most highly integrated silicon photonics product to date: a 100-Gb coherent transceiver in a CFP module for metro applications [17]. Acacia started selling its AC-100 CFP module in late 2014 (Fig. 4.6).

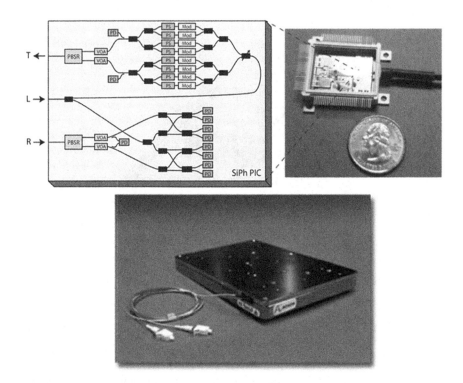

Figure 4.6 Acacia Communications' coherent transceiver. From Acacia Communications.

The company has integrated a variety of optical functions on its silicon photonics transceiver coherent transmission chip.

These include four modulators to implement the modulation scheme for long-haul optical transport, polarization beam splitters to separate the two polarizations of light, a 90-degree optical hybrid whose function is to mix the received signal with the reference signal to implement coherent detection, variable optical attenuators—a component discussed in the Kotura case study—to avoid receiver saturation, and photodetectors to recover the transmission and for monitoring purposes.

In 2015 Acacia introduced its AC-400, the first dual-carrier transceiver [18]. The two-wavelength "superchannel" design supports 200-, 300-, and 400-Gb/s line rates. To achieve the higher data rates, more complex quadrature amplitude modulation schemes are used. Coherent transmission and the associated modulation schemes are discussed in Chapter 5, Metro and Long-Haul Network Growth Demands Exponential Progress.

Acacia sees CMOS-based optics as key to lowering design costs. The company highlights testing the chips while still on the silicon wafer—wafer-level testing—as one important cost-saving measure. And photonic integration reduces the number of discrete components. The laser is the only optical function not integrated on-chip.

Using integration, Acacia avoids the use of lenses, for example. The chip does not need to be hermetically sealed, a common requirement to protect the laser and other optical components from performance degradation due to moisture and dust. The silicon photonics waveguides also confine the light, resulting in a compact design, while integration reduces the touchpoints needed when making the module, reducing manufacturing costs.

Acacia's expertise is not confined to silicon photonics. The company makes its own digital signal processing ASIC for coherent transmission (see Appendix 2: Optical Transmission Techniques for Layer 4 Networks). Combining the two—a photonic integrated circuit (PIC) and the coherent DSP-ASIC—allows Acacia to trade-off the performance of the optics with the compensating electronics to deliver a cost-, size-, and power-optimized coherent CFP optical module.

Acacia faces performance challenges, however. Silicon photonics waveguides have a relatively high insertion loss, and for coherent transmission optical loss is paramount since it dictates reach. In turn, individual components cannot be optimized independently as overall trade-offs must be made. But certain optical performance issues can be countered in the electronic domain.

The coherent transceiver is a complex, highly integrated design. However, the long-distance market uses a relatively low number of such modules: tens to hundreds of thousands of units each year, less than the number of transceivers in a single large data center. This is a drawback, as volumes benefit manufacturing while overall revenues are needed to fund next-generation product development.

4.3.5 Delivering a Performance Edge

A performance edge is a common thread for the four early silicon photonics product case studies. Yet while these are trailblazing products, their performance edge is not decisive and certainly not permanent.

The submicrosecond response time of Mellanox's variable optical attenuator is among the fastest in the industry. Mellanox also claims that competing devices cannot match the cost of its variable optical attenuator; the performance edge has led to volumes that in turn have lowered unit costs. These factors have preserved the company's lead.

Luxtera's PSM4 transceiver could also be implemented using indium phosphide, as could the sharing of the laser across four channels. But Luxtera's 40-Gb PSM4 was the first such device to market, giving it first-mover advantage. Luxtera was also first to ship a 100-Gb PSM4 transceiver, in part because of the expertise it gained at 40 Gb. But now it faces stiff competition from other optical module players.

Cisco gained a year's competitive lead with its 100-Gb CPAK transceiver. But now the market offers 100GBase-LR4 modules using indium phosphide in a CFP2 pluggable module as well as the smaller CFP4 and QSFP28 pluggable modules.

Meanwhile, Acacia's coherent CFP transceiver is a more complex design, both serving and competing with the system vendors. So far the company continues to show solid growth, which was further confirmed with its successful initial public offering in May 2016 [23].

Acacia's coherent modules use an unprecedented degree of integration for a silicon photonics design, yet the design does not match the integration levels achieved using indium phosphide technology. As will be discussed in Chapter 5, Metro and Long-Haul Network Growth Demands Exponential Progress, system vendor company Infinera offers commercially available products based on a PIC implemented in indium phosphide that delivers up to 2.4 Tb of capacity [32].

Longer term, however, we argue Acacia's solution is better suited to deliver compact low-cost modules. Silicon photonics promises higher density as more functions become integrated on silicon, and it offers copackaging opportunities with the coherent DSP-ASIC as well as other electrical parts such as the transimpedance amplifier and the optical transmitter driver circuitry.

Table 4.1 Performance Differentiators of the Four Product Case Studies			
Supplier	Product	Performance Differentiator	Comments
Mellanox (Kotura)	Variable optical attenuator	Fastest response time	Gained and secured a market lead
Luxtera	A PSM4 transceiver at 40 and 100 Gb	Low cost, low power	Only one laser, an expensive part of the bill of materials, is used rather than four
Cisco	The 100GBase-LR4 CPAK transceiver in a small form factor	First to market with the highest density front plate switch	The use of silicon photonics advanced system performance
Acacia	100-Gb transceiver chip for coherent long-distance transmission	First to market with single-chip coherent CFP transceiver	Compact, low-power design lowers the cost of coherent modules

The performances advantages of the case study products are summarized in Table 4.1.

In summary, a performance edge is key for silicon photonics to gain a market foothold. But for further successes, product development must continue and that requires revenues. Silicon photonics must also overcome additional issues to keep developing, as is now discussed.

4.4 WHAT SILICON PHOTONICS NEEDS TO GO MAINSTREAM

The four early product case studies detailed in Section 4.3 are important on many levels. They encourage other silicon photonics players and help build momentum. Product sales also raise revenues, vital for companies not just to balance the books but to fund future product development. Without this, silicon photonics stalls.

However, much work remains to take silicon photonics from early products to the stage where a design and manufacturing infrastructure is in place to serve a global industry.

A much-heard argument for silicon photonics is its potential to reduce component cost by piggybacking on the huge investment already made by the chip industry. There is some truth to this, but as usual things are more complex than they seem.

Silicon photonics may share the chip industry's equipment and manufacturing processes, but it has its own processes and testing requirements.

Companies are also pursuing different designs and techniques for laser attachment, photodetectors, silicon waveguides, modulators, and other optical functions. This diversity, or lack of market convergence, is troublesome because developing manufacturing rigor, uniform tools, and the aggregation of volumes are all impeded. This runs counter to what brought success in the chip industry. That said, the chip industry also started with a diversity of approaches and solutions before settling down.

Silicon photonics is still not mature enough to direct developers to the best solutions covering all aspects of manufacturing. For example:

- New optical circuit probes and testing techniques must be developed. This is photonics after all; electronic integrated circuits have never needed such things.
- Packaging is also a key component of a photonic device's cost, and coupling light to the silicon photonics chip presents challenges, as discussed in Chapter 3, The Long March to a Silicon Photonics Union.
- Eliminating hermetically sealed packages to protect the optical components from dust and moisture is a benefit of silicon photonics since the optical functions are buried in the chip. Nonhermetic packages are needed for all optical functions, yet some vendors still use hermetically packaged lasers, for example.

Device testing and yields are also issues. Wafer testing is being developed by many of the silicon photonics players. Correlating the device performance to the wafer measurements is a key factor in reducing cost. But such correlation comes with experience, once millions of devices have been fabricated.

Getting established foundries to back and offer silicon photonics services is also a critical issue. Freescale, acquired by NXP (itself subject to a bid from chip company, Qualcomm), and STMicroelectronics are active industry participants. Intel has its captive foundry. And Cisco is big enough to dedicate resources to silicon photonics. Yet we worry that there is not enough volume to interest the big foundries while supporting the research and development needed to enable commercialization of silicon photonics. That will likely come, but more likely only when the chip industry needs such technology.

Other elements of optical include eliminating lenses used to focus and collimate light, and the use of isolators to protect the waveguide from spurious optical reflections that affect overall operation. But there is no one uniform approach to packaging even though the industry well understands that it is one of the most important paths to lowering cost.

These processes will evolve just as they have for the semiconductor industry. Production time will decrease and product yields will go up, but it takes time. If silicon photonics is to deliver on its low-cost promise, large volumes will be needed and that requires markets.

The good news, and the subject of Section 4.5, is that 100-Gb links for the data center constitute one such high-volume market. But for now silicon photonics revenues are a fraction of the total sums invested to develop the technology, as is now discussed.

4.4.1 Hefty Investment, Little Return

Table 4.2 shows the investment in silicon photonics start-ups by venture capital firms in the last decade, while Table 4.3 highlights silicon photonics company acquisitions. Summing the venture capital

Table 4.2 Venture Capital Funding for Silicon Photonics Start-Ups Based on Publicly Available Data		
Company	Investments (millions)	Comments
Kotura	$38.7	4 rounds
Rockley Photonics	$13	Estimate and assuming series B
Lightwire	$18	1 round
Ranovus	$35	2 rounds
Skorpios Technologies	$67.9	5 rounds
Luxtera	$91.1	4 rounds
Aurrion	$22.5	4 round
Caliopa	$3.6	2 rounds
Sicoya	$3.8	1 round
Ayar Labs	$2.5	1 round
Total	$296.1	–
Source: Data from Crunchbase, company reports.		

Table 4.3 Silicon Photonics Company Acquisitions		
Acquirer/target	Date	Amount (millions)
Cisco/Lightwire	March 2012	$271
Mellanox/Kotura	August 2013	$82
Huawei/Caliopa	September 2013	$5[a]
Macom/BinOptics	December 2014	$230
Ciena/Teraxion	January 2016	$32
Juniper/Aurrion	August 2016	$165
Total	–	$785

[a]Estimate.
Source: Data from company reports.

funding and the total amount spent on mergers and acquisitions, the investment comes to $1081 million. And this ignores the technology development work done over the last decade by IT giants IBM, Intel, and Hewlett-Packard, among others.

Meanwhile, the total telecom and datacom markets that silicon photonics can address was worth almost $1 billion in 2015. The investment appears large compared to the optical component market size.

Silicon photonics as a new technology must take market share from established photonic technologies in a fiercely competitive marketplace. We estimate that silicon photonics has captured less than 20% market share in two of the most promising early markets for the technology: the variable optical attenuator market and the market for 40- and 100-Gb/s optical interfaces.

Market research firm ElectroniCast estimates the variable optical attenuator market was worth $200 million in 2015 [33]. We estimate Mellanox's variable optical attenuator, with its superior optical performance, has captured half the market, raising $100 million in revenues.

Ovum, another market research firm, valued the 40- and 100-Gb single-mode transceiver market at $800 million in 2015 [34]. These among the most competitive optical component and module markets, making it hugely challenging for an entrant to gain share without a compelling competitive advantage.

Moreover, only recently have silicon photonics 100-Gb/s commercial products become available. We estimate that the silicon photonics

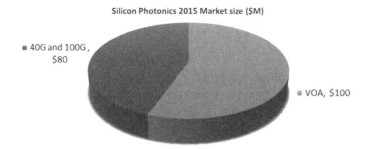

Figure 4.7 Revenues in 2015 from silicon photonics product sales.

market share was 10%, or $80 million, in 2015. Summing the two markets, total silicon photonics revenue in 2015 was $180 million (Fig. 4.7).

For 2016, market research firm LightCounting forecasts that transceiver sales using silicon photonics will reach $300 million. It shows the market is growing, but recall that $1081 million has been spent developing the technology. And that ignores the decade plus of R&D funding by the large IT players.

4.5 100-GB MARKET REVENUES ARE INSUFFICIENT FOR SILICON PHOTONICS

Martin Schell, professor for Optoelectronic Integration at Technical University Berlin, and director of the Fraunhofer Heinrich Hertz Institute, in a 2013 presentation at the European Conference on Optical Communications (ECOC), estimated that $20 million is needed to develop a silicon photonics device using the mature 130-nm CMOS process [35]. This requires a vendor to generate $200 million in sales over a 5-year period to fund its next-generation product development.

The figure is arrived at as follows. The typical operating margin— effectively net profit—of an optical component module vendor is 10%. Given that a product's lifecycle is 5 years commonly, a company needs $200 million in sales to raise the $20 million quoted by Schell. Silicon photonics companies must strive for such self-sufficiency if they are to get beyond this phase of steep investment with little return.

Market research firm Ovum forecasts aggregate revenues of nearly $6.0 billion in 2016–20 for 100-Gb single-mode transceivers with up to 10 km reach in the data center. A second market, the 100-Gb coherent transceivers used in Layer 4, is valued at $3.5 billion through to 2020.

Any silicon photonics vendor that gains a 20% market share approaches self-sufficiency in either of these markets. But only one or two companies in each market can expect to achieve that.

4.5.1 The Near-Term Data Center Opportunity

The silicon photonics data center opportunity is both significant and taxing.

When the large Internet content providers turn up a new data center, as many as 100,000 optical transceivers may be used. The market for 100-Gb/s interfaces in the data center started to ramp in the second half of 2016, and because this opportunity is well recognized by the optical component and module players, competition is intense.

Luxtera, Molex, Cisco, Kaiam, Lumentum, Intel, and Mellanox are all companies that have 100-Gb silicon photonics transceiver products, while IBM and Skorpios are developing silicon photonics transceivers. All these players also compete with transceiver market leaders Finisar, Avago, Fujitsu Optical Components, Sumitomo, Oclaro, and Lumentum, which have indium phosphide−based products. And firms like InnoLight, Applied Optoelectronics Inc., Colorchip, and Source Photonics are other players attacking the 100-Gb opportunity.

There are also multiple 100-Gb interconnect interface types, which fragments market revenue. The main 100-Gb interfaces are the PSM4, 100GBase-LR4, 100GBase-LR4-Lite, CWDM4 and its more stringent specification counterpart, the CLR4, and OpenOptics [36].

Revenues for transceivers inside the data center are attractive. However, no one solution addresses all requirements due to different reaches, multisource agreements, and standards. This means multiple products need to be developed, increasing the overall investment companies must make. And with many suppliers, some with deep pockets and others with formidable technology, the competition limits the market share any one company can expect.

Fig. 4.8 illustrates the challenges. It shows estimated revenues by technology for single-mode 100-Gb transceivers up to 10 km. As the

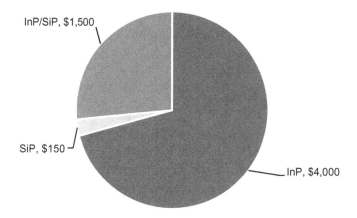

InP/SiP, $1,500

SiP, $150

InP, $4,000

Figure 4.8 100-Gb single-mode aggregate transceiver revenues by technology for 2016–20 ($ millions).

incumbent technology for 100GBase-LR4, the de facto standard at this reach, indium phosphide is best positioned to win the lion's share of the revenue. The first single-mode 100-Gb transceiver commercialized for reaches to 10 km is based on indium phosphide, and while it is not expected to be the highest volume product, its tight specifications drive higher prices, resulting in the biggest revenue among its interconnect peers.

The PSM4 and OpenOptics are expected to be supported by silicon photonics, but the CWDM4 and CLR4 will use either silicon photonics or indium phosphide technologies.

A market leader usually emerges when it becomes the first to introduce a new product, gaining over half the market share. Finisar's 100GBase-LR4 transceiver in the CFP form factor is one such example. A company needs to develop products that win market share, yet choosing which 100-Gb products to back is not straightforward.

What does all this mean?

The 100-Gb data center opportunity is timely for silicon photonics, but this market is insufficient to deliver self-sustaining revenues for transceiver suppliers. Venture capital, public markets, and even government resources will need to be tapped to help take suppliers to the next level. There is also another development: system vendors are acquiring silicon photonics players, as we discuss in Section 4.6.

4.5.2 The Coherent Transceiver Market Has Its Own Challenges

Acacia is the only vendor currently shipping a silicon photonics–based coherent transceiver. Fujitsu Optical Components and Oclaro are shipping competing products that deliver 100- and 200-Gb coherent links in a CFP2 Analog Coherent Optics module, and they are being joined by Finisar and Lumentum.

But the competitive landscape here is different to that of the 100-Gb data center market because of the digital signal processor ASIC and the electronics needed to sample and convert the analog signal to a digital one. These are critical technologies for product differentiation.

Optical equipment vendors such as Huawei, Infinera, Ciena, Nokia, and Cisco all have proprietary digital signal processor ASICs owing to the device's importance in determining overall optical link performance. The DSP-ASICs are becoming highly sophisticated in the features they enable and give companies an important time-to-market advantage [37,38]. Meanwhile, merchant DSP-ASICs are available from suppliers such as ClariPhy (set to be acquired by Inphi) and NEL, but the optical equipment vendors' designs are ahead of the game.

Acacia's ambitious design highlights the merits of silicon photonics. Its AC-400 transceiver supports 200-, 300-, and 400-Gb rates, making it the market's first flexible-rate compact module that also has low power consumption. Silicon photonics is important in realizing this performance. But other companies are developing silicon photonics products to compete with Acacia. NTT, e.g., has detailed development results for such a design.

4.5.3 Emerging Opportunities for Silicon Photonics

The focus of this chapter has been the 100-Gb/s opportunity, but higher data rates are emerging in the data center, such as 400-Gb Ethernet. Four formats are being standardized by the IEEE for 400-Gb Ethernet which uses more sophisticated modulation than standard 100-Gb technologies. The four styles of 400-Gb Ethernet also require more lanes to achieve the higher data rate [39]. Increasing the number of lanes favors photonic integration and hence silicon photonics. Rather than coming relatively late as happened at 100 Gb, silicon photonics will be competing with indium phosphide technology as the

400-Gb standard emerges. Silicon photonics will therefore be an even more formidable competitor at this rate.

There are also new developments for optical transport suited to silicon photonics: higher-speed coherent optics, and new direct detection technologies for shorter-reach (up to 130 km) data center interconnect. These are discussed in Chapter 5, Metro and Long-Haul Network Growth Demands Exponential Progress.

Another trend is to move optics away from the faceplate and closer to the silicon—first on-board optics and then copackaged optics; developments explored in Chapter 7, Data Center Architectures and Opportunities for Silicon Photonics. These applications are the most significant opportunities for silicon photonics in the datacom and telecom domains.

4.6 THE SILICON PHOTONICS ECOSYSTEM: A STATE-OF-THE-INDUSTRY REPORT

This chapter has highlighted two key findings:

- The successful silicon photonics products to date have all offered a particular performance advantage compared to the established products. This may be an optical performance edge, a first-mover advantage, or an advantage at the system level.
- The emerging 100-Gb interface market for the data center offers a much needed high-volume opportunity, yet this market will be insufficient to deliver self-sustaining revenues for the transceiver suppliers.

In the tech sector, acquisitions offer one exit strategy for investors and start-ups, and usually help ensure the technology's long-term survival. The trend of system vendors acquiring silicon photonics suppliers that was first evident in 2012–14 has reemerged in 2016. Ciena, Cisco Systems, Huawei, Juniper Networks, and Mellanox have all made such a move.

This shows that system vendors recognize that silicon photonics is a technology they need for their future product plans and, more importantly, they cannot just work with and depend on silicon photonics or chip players; it is a technology they must own.

Another development that bodes well for silicon photonics is the emergence of a strong ecosystem.

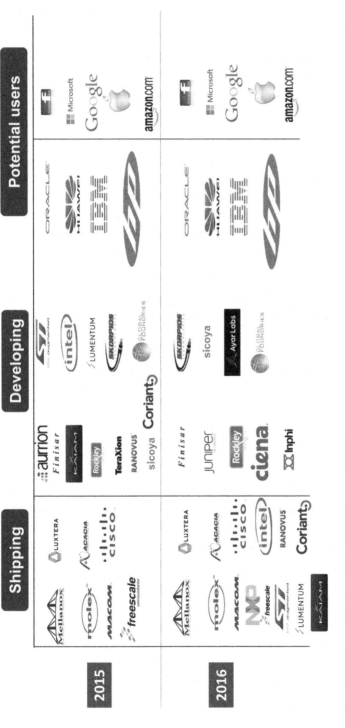

Figure 4.9 Vendors participating in silicon photonics' commercialization.

Fig. 4.9 shows a 2015 snapshot of market participants and how it has evolved in 1 year. Represented are diverse elements of the market: start-ups and established vendors; system, module, and component vendors; incumbent transceiver vendors; and the equipment makers. Also included are end customers as well as suppliers making products for various market segments, including the Layer 4 telecom network core and Layer 3 data centers.

Also featured are multiple top-ten global chip companies. This chapter has focused on the current telecom and datacom market opportunities and rightly so, as this is the current market route for silicon photonics. But we expect electronic designers will also require silicon photonics for their chip designs, and when that happens the large semiconductor foundries will start to lead the market. This will be a disruptive development, a topic explored in Chapter 8, The Likely Course of Silicon Photonics.

Key Takeaways

- Silicon photonics is moving toward general market adoption, having passed the phases of excessive expectation and deep disillusion. However, much work remains to achieve widespread adoption.
- Several early successful silicon photonic products have all exceeded the performance of established technologies. These advantages include improved or unmatched performance, novel designs, a time-to-market advantage, and advanced integration.
- Silicon photonics must capture enough market share and revenues to pay for next-generation product development. We estimate that each market must generate revenues of $200 million over a 5-year period for a company if this is to happen.
- The emerging 100-Gb transceiver opportunity connecting data center equipment promises huge revenues. But the volumes and revenues silicon photonics will gain will be limited due to market fragmentation, in terms of both interfaces and technologies. As such, the 100-Gb market for the data center alone will not bring the technology to mainstream status.
- Laser attach, fibering the chip, testing, packaging, common manufacturing processes, and widespread volume design and foundry services are all challenges the industry must tackle if silicon photonics is to fulfill its potential.
- Emerging developments such as high-speed coherent optical transmission, 400-Gb Ethernet, and moving optics closer to silicon will also benefit silicon photonics, as will the silicon photonics ecosystem that is taking shape.

- System vendors continue to acquire silicon photonics players. They recognize they must own the technology for their product roadmaps. This is another development ensuring the technology's long-term future.

REFERENCES

[1] Arthur WB. The nature of technology: what it is and how it evolves. New York: Free Press; 2009.

[2] The quiet period of silicon photonics. Gazettabyte, < http://www.gazettabyte.com/home/2015/8/12/the-quiet-period-of-silicon-photonics.html > ; August 12, 2015.

[3] Fenn J, Raskino M. Mastering the hype cycle: how to choose the right innovation at the right time. Boston, MA: Harvard Business Press; 2008.

[4] Shoji T, Tsuchizawa T, Watanabe T, Yamada K, Morita H. Low loss mode size converter from 0.3 μm square Si wire waveguides to single mode fibres. Electron Lett 2002;38:1669−70.

[5] Almeida VR, Panepucci RR, Lipson M. Nanotaper for compact mode conversion. Optics Lett 2003;28(15):1302−4.

[6] Intel silicon photonics research, < http://www.intel.com/content/www/us/en/research/intel-labs-silicon-photonics-research.html > .

[7] Mario Paniccia: We are just at the beginning. Gazettabyte, < http://www.gazettabyte.com/home/2016/5/23/mario-paniccia-we-are-just-at-the-beginning.html > ; May 23, 2016.

[8] Cisco announces intent to acquire lightwire, press release, < https://newsroom.cisco.com/press-release-content?articleId=675179 > ; February 24, 2012.

[9] Mellanox Technologies Ltd. announces definitive agreement to Acquire Kotura, Inc, press release, < http://ir.mellanox.com/releasedetail.cfm?ReleaseID=765188 > ; May 15, 2013.

[10] Molex Purchases Luxtera's silicon photonics-based active optical cable (AOC) business; Partners on Future AOC Development, press release, < http://www.molex.com/mx_upload/editorial/894/20110111_Luxtera_AOC.html > ; January 11, 2011.

[11] Industry note: silicon photonics: data center disruptor, James Kisner. Jeffries Equity Research; February 22, 2013.

[12] Huawei completes acquisition of Caliopa, press release, < http://caliopa.com/news-and-events/article/huawei-completes-acquisition-of-caliopa > ; September 9, 2013.

[13] Macom announces definitive agreement to acquire BinOptics Corporation, press release, < http://ir.macom.com/releasedetail.cfm?releaseid=883689 > ; November 18, 2014.

[14] US Securities and Exchange Commission. Macom's 10-Q form for the quarter ending, < https://www.sec.gov/Archives/edgar/data/1493594/000119312515029663/d850710d10q.htm > ; January 2, 2015.

[15] Intel's disaggregated server rack, research note. Moor Insight and Strategy, < http://www.moorinsightsstrategy.com/research-note-intels-disaggregated-server-rack/ > .

[16] Cisco CPAK for 100-Gbps solutions, < http://www.cisco.com/c/en/us/products/collateral/optical-networking/ons-15454-series-multiservice-provisioning-platforms/white_paper_c11-727398_031813.html > .

[17] Acacia uses silicon photonics for its 100G coherent CFP. Gazettabyte, < http://www.gazettabyte.com/home/2014/3/18/acacia-uses-silicon-photonics-for-its-100g-coherent-cfp.html > ; March 18, 2014.

[18] Acacia unveils 400 Gigabit coherent transceiver. Gazettabyte, < http://www.gazettabyte.com/home/2015/3/31/acacia-unveils-400-gigabit-coherent-transceiver.html >; March 31, 2015.

[19] Intel delays part for high-speed silicon photonic networking. Computerworld, < http://www.computerworld.com/article/2879077/servers/intel-delays-part-for-high-speed-silicon-photonic-networking.html >; February 3, 2015.

[20] IBM demos a 100 gigabit silicon photonics transceiver. Gazettabyte, < http://www.gazettabyte.com/home/2015/7/10/ibm-demos-a-100-gigabit-silicon-photonics-transceiver.html >; July 10, 2015.

[21] Ciena shops for photonic technology for line-side edge. Gazettabyte, < http://www.gazettabyte.com/home/2016/1/25/ciena-shops-for-photonic-technology-for-line-side-edge.html >; January 25, 2016.

[22] Juniper networks to acquire Aurrion for $165 million. Gazettabyte, < http://www.gazettabyte.com/home/2016/8/8/juniper-networks-to-acquire-aurrion-for-165-million.html >; August 8, 2016.

[23] Acacia soars 35 percent in second tech IPO of the year. Techcrunch, < https://techcrunch.com/2016/05/13/acacia-soars-36-in-second-tech-ipo-of-the-year/ >; May 13, 2016.

[24] Macom announces industry's first CWDM4 L-PICTM for 100G datacenter applications, press release, < http://files.shareholder.com/downloads/AMDA-G79D4/0x0x880034/DD018906-ACA5-4806-ACFA-822AB9276ABE/MTSI_News_2016_3_8_General_Releases.pdf >; March 8, 2016.

[25] KAIAM demonstrates world's first complete 100 Gb/s CWDM4 silicon photonics transceiver at OFC 2016, press release, < http://kaiamcorp.com/?page_id=10019 >; March 17, 2016.

[26] Lumentum highlights new technologies and solutions at OFC 2016, press release, < https://www.lumentum.com/en/media-room/news-releases/lumentum-highlights-new-technologies-and-solutions-ofc-2016 >; March 18, 2016.

[27] Ranovus shows 200-gigabit direct detection at ECOC. Gazettabyte, < http://www.gazettabyte.com/home/2016/9/20/ranovus-shows-200-gigabit-direct-detection-at-ecoc.html >; September 20, 2016.

[28] Intel's 100-gigabit silicon photonics move. Gazettabyte, < http://www.gazettabyte.com/home/2016/8/21/intels-100-gigabit-silicon-photonics-move.html >; August 21, 2016.

[29] Luxtera introduces industry's lowest power 40G AOC, press release, < http://12.162.78.22:8080/luxtera/20091116LuxteraLowPowerBlazar.pdf >; November 16, 2009.

[30] The case for silicon photonics. Optical Connections, Issue 5, Q3 2015, pp. 16−17. < http://opticalconnectionsnews.com/wp-content/uploads/2015/06/Optical-Connections-2015-Q3-44pp-Final-LR.pdf >.

[31] Luxtera ships one millionth silicon photonic transceiver product, press release, < http://www.marketwired.com/press-release/update-luxtera-ships-one-millionth-silicon-photonic-transceiver-product-2159608.htm >; September 19, 2016.

[32] Infinera goes multi-terabit with its latest photonic IC, < http://www.gazettabyte.com/home/2016/3/30/infinera-goes-multi-terabit-with-its-latest-photonic-ic.html >; March 30, 2016.

[33] ElectroniCast: fiber-optic attenuator market worth $370M in 2018, Published in LightWave, < http://www.lightwaveonline.com/articles/2013/11/electronicast-fiber-optic-attenuator-market-worth-370m-in-2018.html >; November 25, 2013.

[34] Ovum forecast, Total Optical Component Forecast Spreadsheet: 2014−2020, TE0017-000051; August 2015.

[35] Martin Schell, Economics of PICs, ECOC 2013 WS2 Low-cost access to PICs.

[36] The OpenOptics MSA, < http://www.openopticsmsa.org >.

[37] Next-generation coherent adds sub-carriers to capabilities, < http://www.gazettabyte.com/home/2016/1/24/next-generation-coherent-adds-sub-carriers-to-capabilities.html > ;   January 24, 2016.

[38] Nokia's PSE-2 delivers 400-gigabit on a wavelength. Gazettabyte, < http://www.gazettabyte.com/home/2016/6/8/nokias-pse-2s-delivers-400-gigabit-on-a-wavelength.html > ; June 8, 2016.

[39] The IEEE P802.3bs 400 gigabit per second Ethernet Task Force, < http://www.ieee802.org/3/bs/ > .

Metro and Long-Haul Network Growth Demands Exponential Progress

Growth is what matters. People don't take big risks and do interesting things to attack flat or contracting markets.

Andrew Schmitt [1]

With the relentless growth in traffic, telecom needs something new. You can't just take what we have and continue to cost-reduce.

Karen Liu [2]

5.1 THE CHANGING NATURE OF TELECOM

The first transcontinental telephone call was made just over a century ago. The conversation took place between inventor Alexander Graham Bell in New York City and his San Francisco–based former assistant, Thomas Augustus Watson [3].

One hundred years later, voice remains a core telecom service, although it is now a stream of 1s and 0s. And in this new digital world, the world's leading communications service providers (CSPs) or telcos—the likes of AT&T, China Mobile, Deutsche Telekom, and NTT—offer much more than voice. The telcos deliver entertainment services such as streamed video and music as well as business services distributed over their sophisticated fixed and mobile networks. Indeed, perhaps the telcos' most prized subscriber offering is not so much a service as a piece of hardware: the smartphone, as powerful as a laptop computer yet small enough to be inseparable from its user.

Telcos recognize the importance of owning the connection linking a subscriber to their networks. Their strategy is to secure a subscriber via a handset or a broadband residential gateway and then sell them services. But such connectivity is also central to the businesses of the Internet content providers, such as Google, Apple, Facebook,

Silicon Photonics. DOI: http://dx.doi.org/10.1016/B978-0-12-802975-6.00005-3
Copyright © 2017 Daryl Inniss and Roy Rubenstein. Published by Elsevier Inc. All rights reserved.

Microsoft, and Amazon, that deliver "over-the-top" services over those same telcos' networks.

Fixed and mobile telcos dominate the service providers' revenues, based on data from market research firm, Ovum, as shown in Fig. 5.1. But their revenues are largely flat and are expected to remain so through 2020. There are good reasons for this: the telecommunications industry is fiercely competitive, markets are regulated heavily, and there is a limit as to how much a subscriber or household will pay each month for telecom services.

The ability to deliver services over a secure network to millions of users at home, at work, and when on the move remains a key telco strength. And telcos are seeking new opportunities to grow their businesses. These include investing in TV, video, and content [4], cloud services [5], Internet businesses [6] and, in the case of Softbank, acquiring the leading chip player, ARM Holdings [7]. Another development that promises to have a huge impact across many industries is the Internet of Things, an emerging market that will expand greatly the number of machines linked to the Internet.

The Internet content providers, in contrast to the telcos, are much younger companies and are experiencing sharp growth with their massive customer numbers and targeted advertising, the subject of Chapter 6, The Data Center: A Central Cog in the Digital Economy. Their growing revenues and continual striving to advance their businesses using datacom and telecom technologies are an attractive lure

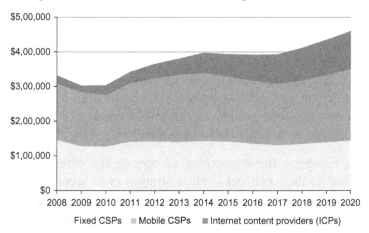

Figure 5.1 Global service provider revenue forecast (millions), 2008–20. Courtesy of Ovum.

for equipment makers and optical module and component companies. As analyst Andrew Schmitt notes, technology firms do not take big risks and do interesting things to attack flat or contracting markets [1].

Fig. 5.2 shows the capital expenditures of the different types of service providers, forecasted through to 2020. Capital expenditure, or capex, refers to how much service providers spend on equipment and premises. Mobile CSPs—wireless carriers such as Verizon Wireless and T-Mobile—also include buying radio spectrum as part of their capex.

Telcos also account for the bulk of the overall providers' capex spending, as shown in Fig. 5.2. Approximately 10% of the telcos' capex is spent on optical networking. This makes telcos the dominant customers for optical communications, having overseen the critical technological steps that have given rise to cost-effective, wavelength-division-multiplexed optical transmission. The telcos' involvement is not happenstance; their businesses depend on such advances.

Meanwhile, the strong growth of the Internet content providers, forecast to continue in the coming years, means their spending cannot be ignored. The Internet businesses are experiencing rapid growth in traffic, driving their demand for networking capacity. And that is leading them to drive new requirements for optical systems and components.

Equipment vendors are responding to the Internet content providers' demands with optical transmission gear tailored to their needs.

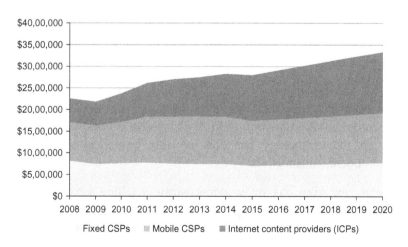

Figure 5.2 Global service provider capital expenditure forecast (millions), 2008–20. Ovum.

This is a significant development. It is the first time in recent history that custom optical equipment has been made for an end customer other than a telco—an equipment category known as data center interconnect and discussed in Section 5.4.

The continual growth in traffic is presenting telcos and Internet content providers with a new challenge: the relentless growth in traffic will cause transport costs to finally rise.

Similar to Moore's law, network operators have experienced a continual reduction in the cost of optical transport, measured in cost per transported bit. But network traffic growth is filling up fibers—for years seen as having boundless capacity—and this will raise the cost of optical transport. Internet content providers, with their large-scale data centers, are experiencing the fastest growth in traffic and will meet this challenge first, as discussed in Section 5.2.

The impending rise in transport costs will have important consequences:

- The Internet content providers and the telcos will do everything they can to fully use their existing installed fiber.
- Optical transport equipment vendors will focus on cost-reducing their platforms. Systems can already fill a fiber with data. The next step will be to make such systems more compact, more power-efficient, and cheaper.

Both developments will require innovation at the systems, optical module, and component levels. Optical integration—whether in the form of smaller pluggable modules, embedded modules, new superchannel transponder designs, and even space-division multiplexing—will all play a role.

This chapter highlights how silicon photonics has already established itself as a design alternative to indium phosphide, a notable achievement given that the technology has only recently been deployed for long-distance optical transport.

Silicon photonics has yet to demonstrate a telling advantage. Line-side optics volumes are relatively modest, making it harder for silicon photonics to differentiate itself. But the advancement in signaling schemes and the close interworking of optics and electronics, coupled with the trend to develop ever more compact platforms, indicate opportunities for silicon photonics.

In turn, leading optical transport vendors now have silicon photonics technology in-house. And it is the systems vendors, with their larger investment budgets and need for product differentiation, that will likely advance silicon photonics for optical transport more than the optical component and optical module makers.

5.2 INTERNET BUSINESSES HAVE THE FASTEST NETWORK TRAFFIC GROWTH

Internet traffic carried by the telcos' networks is growing exponentially each year, at an estimated average rate of 20–30%. There are several reasons for such growth: more subscribers and machines are connecting to the network each year, and the nature of the interactions is evolving. Enterprises' data is increasingly being stored remotely—in the cloud—while the use of video, a particularly demanding service in terms of network capacity, is growing.

Although the telcos' 20–30% annual growth is significant, it turns out to be at the modest end of the spectrum. Cable operators are experiencing 60% growth annually, whereas the Internet content providers are seeing 80–100% growth year-on-year [8]. With an annual near-doubling of traffic, it does not take many years to stretch the capabilities of optical transport systems and the network.

In the early 2000s, typical dense wavelength-division multiplexing systems supported up to 80 10-Gb wavelengths, or 0.8 Tb. Nortel Networks, then the leader in such systems, first announced in 1999 its OPTera 1600G system that supported as many as 160 10-Gb wavelengths, a huge capacity at the time. But this was during the optical boom and bandwidth exuberance; no telco needed such capacity.

Today's long-haul systems support 96 100-Gb wavelengths, nearly 10 Tb of capacity overall. And state-of-the-art systems implementing advanced modulation schemes that support 200 Gb on an optical carrier or wavelength further boost the system's optical transport capacity to 20–25 Tb, albeit over shorter link distances than systems operating at 100 Gb/s. Nokia's PSE-2-based systems can support up to 35 Tb over a fiber in the C band but only when the distances are short: 100–150 km [9]. Nokia also announced that its system will support the L band that will double the capacity to 70 Tb: 35 Tb across the C band and 35 Tb across the L band [10].

The optical industry is thus rising to the challenge of accommodating the exponential growth of Internet traffic. But hurdles remain. Commercial systems are approaching the upper limit of how much can be carried on a fiber, known as the nonlinear Shannon limit [11,12], for transmission distances of hundreds and thousands of kilometers.

The first thing to note about the nonlinear Shannon limit is that the traffic-carrying capacity of a fiber varies with distance, as shown in Fig. 5.3. Much more data can be sent over shorter spans of a few hundred kilometers than over distances of several thousands of kilometers.

The y-axis of the Fig. 5.3 graph is a measure of the efficiency of the transported data in terms of how well a transport system is using the fiber's available bandwidth. This is referred to as the spectral efficiency and is measured in bits/s/Hz. Note the nonlinear Shannon limit defines the fiber's maximum spectral efficiency possible for any given distance.

Lastly, Fig. 5.3 also shows the upper boundary of the capacity-filling performance achieved by leading-edge optical systems demonstrated in research laboratories, confirming that systems are already doing an excellent job in exploiting the capacity of fiber.

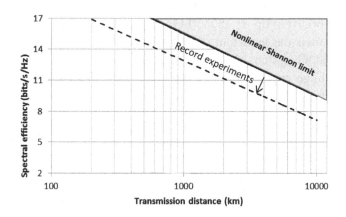

Figure 5.3 How the nonlinear Shannon limit of fiber varies with distance. Based on information from Bell Labs.

Current commercial optical transport systems transport 100-Gb over a 50-GHz wide channel, a spectral efficiency of 2 bits/s/Hz. Fig. 5.3 shows how distances in excess of 10,000 km can be supported using such a scheme.

The need to transport more traffic has led to the adoption of higher transmission rates and higher spectral efficiencies. But these benefits come at the expense of transmission distance. Note, this is a relatively new problem; previous generations of optical transport systems used transmission schemes with a spectral efficiency well below the nonlinear Shannon limit.

Designers are working to reduce the cost of transmission by better using the available capacity. This involves using more complex signaling schemes to effectively increase the spectral efficiency. But because much of the existing capacity is already being exploited, the scope for cost reduction of future systems is limited. This is the significance of optical transport systems approaching the nonlinear Shannon limit.

5.3 THE MARKET SHOULD EXPECT COST-PER-TRANSMITTED-BIT TO RISE

Once a fiber's transmission band reaches its data capacity-carrying limit, the cost hikes up when transmitting the very next bit. That is because the two options available to network planners today involve upfront investment when adding more link capacity.

The planners can either light a separate fiber, or they can use an additional part of the fiber's spectrum—the L band—alongside the C band, the spectral band used currently for optical transport based on dense wavelength-division multiplexing. However, both cases increase the cost-per-bit, a scenario not experienced during the recent history of optical transmission.

Fig. 5.4 summarizes the impact of the per-bit transmission capital cost as data rates increase to meet growing bandwidth requirements.

To be able to compare the relative costs of the different transmission approaches that have been adopted, several assumptions are made. The first is that the optical transport systems are fully loaded with line cards used for optical transmission.

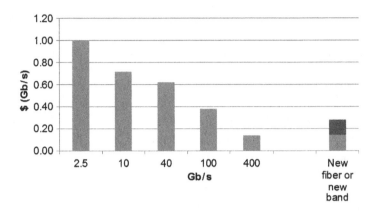

Figure 5.4 Relative transmission cost by data rate. Ovum.

The analysis compares the cost of a 1000-km link and assumes optical amplifiers are spaced every 80 km along the link. Other elements used include multidegree reconfigurable add-drop multiplexers (ROADMs) that support newer 100- and 400-Gb wavelengths or lightpaths, simpler ROADMs for 10- and 40-Gb lightpaths, and fixed optical add-drop multiplexers for the older 2.5-Gb wavelengths. A ROADM is a network element used to switch optical wavelengths between fibers as well as to add new wavelengths or drop wavelengths at a network hub site. A multidegree ROADM allows for such lightpaths to be switched between multiple fiber pairs in a flexible way.

Other assumptions include the use of 40 channels for the 2.5-Gb network, 88 channels at 10 Gb, and 96 channels for the rest. Cost is based on 2015 prices and normalized with respect to 2.5 Gb. The data comes from Ovum's 2015 Optical Networking forecast [13].

The ways used to increase fiber capacity are data rate, the spectral width a wavelength occupies, and the number of wavelength-division-multiplexed channels used. At 2.5 Gb/s, the spectral width is 100 GHz and the maximum capacity is 40 channels, for a total capacity of 100 Gb.

Using 10-Gb lightpaths presents technical challenges. Simply put, the wavelengths are dispersed as they travel across the fiber, and dispersion compensators—extra components—are needed, thereby increasing equipment cost. But the increased data rate, narrower (50 GHz) spectral width, and the network's ability to transmit 88 wavelength-division-multiplexed channels results in a per-bit cost that is lower than at 2.5-Gb/s transmission.

The deployment of the 40-Gb data rate was short-lived between 2005 and 2009, before it started to be superseded by 100-Gb lightpath deployments, but it still drove down the per-bit cost. Systems supported 96 channels at 50-GHz spacing for transmission in the C band. The latest network deployments, using 100- and 200-Gb lightpaths, have a cost based on multidegree ROADMs and 96 channels in the C band.

Once a network planner needs to transmit more data than a fiber's C band can support, they must either light a new fiber or transmit in the L band alongside the C band. Both scenarios increase the cost of transmission. Both capital and operational costs increase if current technology is used, since the operator is effectively running two networks to accommodate traffic growth. The cost increases because all the common equipment—the amplifiers and the ROADMs—must be duplicated to support the extra traffic. And extra equipment is needed at both ends of the fiber.

To continue reducing transmission cost, what is needed is to integrate common equipment so that it can be shared among multiple transmission bands. To reduce operational costs, transmission equipment and the space they occupy and power they consume must come down. This implies that greater integration will be needed to realize a reduced transmission cost. This is where silicon photonics can play a role.

5.4 DATA CENTER INTERCONNECT EQUIPMENT

Section 5.1 mentioned the significance of the Internet content providers in terms of their spending on equipment and how this has resulted in optical equipment vendors developing data center interconnect platforms tailored to those providers' needs. These data center interconnect platforms highlight several important trends:

- Optical transport equipment is becoming denser in terms of the transmission capacity that can be fitted in a box, and this trend will continue.
- A single, albeit stackable, platform is now capable of filling several fibers' worth of capacity.
- Data center interconnect vendors are already offering platforms that use both the C and the L bands to expand dense wavelength-division-multiplexed optical transport.
- Silicon photonics is now being used to link data centers.

• Internet content providers are looking for even cheaper solutions than that offered by the relatively new data center interconnect platforms.

5.4.1 New Requirements and New Optical Platform Form Factors

Data center interconnect platforms are a recent development; Infinera was first to market in 2014 with its Cloud Xpress platform [14]. Other optical transport vendors were already offering equipment that was being used to connect data centers, but Infinera's Cloud Xpress was the first example of a style of platform tailored to the data center that differed from traditional telecom equipment, as is now explained.

Data center managers want to connect sites using point-to-point links, whereas the telcos' Layer 4 networks carry a variety of traffic types and services between many locations, requiring a multipoint-to-multipoint network topology.

Data center managers thus want equipment that can send lots of data without needing the common equipment technologies telcos require. Common equipment such as ROADMs is not needed when the link is point-to-point, and optical amplification is not needed if the data centers link distances are short enough. Other attributes required for the data center include power consumption efficiencies and compactness—conserving power and floor space are key operational expenses that concern data center managers.

Data center managers are used to platforms that scale by simply adding and stacking more cards in slim boxes in a rack. Such cards are referred to as "pizza boxes" (some deep pan, some thin-based) due to their dimensions. This is how data center servers are designed, and it is how managers want their optical transmission equipment.

Cisco Systems' NCS 2015 chassis-based optical transport platform for telcos, and the NCS 1002, its first data center interconnect platform, highlight these differences. The NCS 2015 chassis has 15 card slots, each supporting 200 Gb of traffic. The total 3 Tb of optical transport capacity require a total of 12 rack units, a rack unit being a universal unit of measure to determine the height consumed by a box or stacked platform. In contrast, Cisco's NCS 1002 data center interconnect product crams 2 Tb of line-side capacity into just

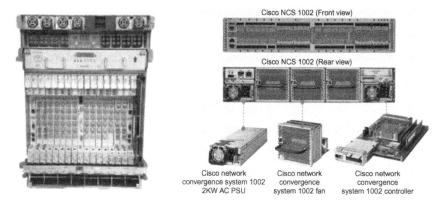

Figure 5.5 Cisco Systems' NCS 2015 and NCS 1002 platforms. From Cisco.

two rack units, improving capacity density fourfold compared to its telco chassis.

Fig. 5.5 shows the Cisco NCS 2015 chassis and NCS 1002 data center interconnect platforms.

5.4.2 From Cloud Xpress to Cloud Xpress 2

It is informative to look at how other optical equipment makers have responded with their data center interconnect platform designs.

Infinera's first-to-market Cloud Xpress platform uses the vendor's own 500-Gb photonic integrated circuit (PIC) chipset in a 2-rack-unit (2RU) box. These boxes can be stacked 16 high to create an 8-Tb line-side capacity optical transport platform [14].

Ciena introduced its Waveserver, which supports 400-Gb line-side capacity using its own optical design and coherent digital signal processing chips, known as DSP-ASICs [15]. Coherent optics is discussed in Section 5.5.1 and in Appendix 2, Optical Transmission Techniques for Layer 4 Networks. The 400 Gb/s is achieved using two carrier signals, each carrying 200 Gb/s. The Waveserver achieves 200 Gb per wavelength by using a more advanced modulation scheme than Infinera's first-generation Cloud Xpress.

The complete platform accommodates up to 44 Waveserver stackable units, yielding 88 wavelengths, a total capacity of 17.6 Tb. Ciena says it can achieve more—up to 25 Tb of capacity—by creating narrower channels using its coherent DSP-ASIC to shape the optical pulses before transmission.

ADVA Optical Networking has advanced data center interconnect further with its CloudConnect platform, which comes in different configurations [16]. All the CloudConnect platforms use a QuadFlex card whose line-side interface supports 100-Gb ultralong-haul to 400-Gb metro-regional transmission, in increments of 100 Gb. Two carriers are used, but different modulation schemes are implemented to achieve the 200-, 300-, and 400-Gb line rates.

The CloudConnect is notable in that it can double total line-side capacity by also using the L band. Using two platforms, a total of 51.2-Tb line-side capacity can be achieved: 25.6 Tb across each of the bands.

Yet another platform, Coriant's Groove G30, hosts eight CFP2-Analog Coherent Optics (CFP2-ACO) pluggable optical modules on a 1-rack-unit card. Each CFP2-ACO supports 100, 150, and 200 Gb using different modulation schemes. Up to 128 wavelengths fit in the C band for a total capacity of 25.6 Tb [17].

A single G30 platform supports 42 cards, resulting in a total line-side capacity of 67.2 Tb. This exceeds ADVA's two-platform capacity, but three fiber pairs are needed to carry the full 67-Tb capacity on a single platform.

In effect, vendors can now design pizza boxes with terabit line-side capacities and stack them so high in one platform that multiple fibers are needed to benefit from all the line-side capacity.

Cisco Systems' NCS 1002 platform supports 250 Gb/s on a single wavelength, more than the other vendors' 200 Gb. This reduces the number of line-side pluggable modules needed on the equipment's faceplate: for Cisco's platform, four 250-Gb CFP2-ACO optical modules can deliver a terabit of capacity instead of five 200-Gb CFP2-ACOs. Overall, Cisco uses 96 wavelengths instead of 120 to deliver a total platform capacity of 24 Tb [18].

Infinera has since launched its second-generation Cloud Xpress 2. It is the vendor's first commercial platform to use its latest-generation PIC and DSP-ASIC that support 1.2 Tb of capacity [19].

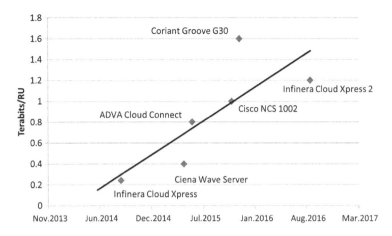

Figure 5.6 Data center interconnect introduction dates and per-rack-unit capacity.

Fig. 5.6 summarizes the data center interconnect platforms discussed in terms of their line-side capacity normalized to a single rack unit and plotted over time.

What is evident is that platform density is increasing, a trend that will continue. Moreover, the two highest density platforms, Coriant's Groove G30 and Infinera's Cloud Xpress 2, are based on PICs, with the highest capacity platform using silicon photonics.

5.5 THE ROLE OF SILICON PHOTONICS FOR DATA CENTER INTERCONNECT

The platforms shown in Fig. 5.6 use various line-side photonics approaches. These include PIC technology, custom line-side optics designed on a line card, and pluggable optical modules in the form of the CFP2-ACO. With the CFP2-ACO, the module houses the optics used for coherent transmission while the accompanying DSP-ASIC resides on the line card.

Infinera's platform uses its 1.2-Tb PIC: six wavelengths each supporting 200 Gb/s of traffic. Note that this is not the maximum capacity of Infinera's latest PIC technology: it has a DSP-ASIC and PIC combination—its Infinite Capacity Engine—that can support up to 12 channels, each at 200 Gb [20,21]. However, the company is using a trimmed-down six-wavelength version for its Cloud Xpress 2.

In contrast, Coriant's G30 uses eight CFP2-ACOs, which when operated at 200 Gb/s equates to a line-side density of 1.6 Tb per rack unit. Coriant says that it has its own silicon photonics technology while also collaborating with strategic partners. The company says that its platform is using both silicon photonics and indium phosphide CFP2-ACO pluggable modules.

What can be concluded is that silicon photonics is already competitive for data center interconnect applications in terms of reach and capacity density but is up against the incumbent technology of indium phosphide.

Infinera's latest-generation indium phosphide PIC was 4 years in development. Used in the Cloud Xpress 2, Infinera has increased line-side density nearly fivefold compared to its first-generation product. Meanwhile, both indium phosphide and silicon photonics designs are being used for the CFP2-ACO optical module in Coriant's platform.

Infinera is unique in that it has reserved its indium phosphide PIC expertise for use in its own platforms; it does not sell its integrated photonics chips to third parties. Selling to third parties is important for silicon photonics vendors if they want to be profitable, given the relatively low volumes associated with line-side interfaces. But several optical transport vendors—Ciena, Coriant, Huawei, Cisco Systems, and Nokia—have their own silicon photonics expertise and can develop their own solutions for the data center interconnect market.

Currently the most integrated silicon photonics design is Acacia's single-chip transceiver that supports up to 200 Gb. This is only a sixth of the capacity of Infinera's newest PIC in the Cloud Xpress 2, although Acacia's design integrates the transmitter and receiver on one chip while Infinera uses separate transmitter and receiver PICs. And as of this writing, it has been 18 months since Acacia announced its chip; one can assume the company is well advanced in developing its next-generation photonic integrated chip design, its next product with even more advanced features.

Meanwhile, the next line-side pluggable optical module development after the CFP2-ACO will be the CFP8-ACO, a module that is flatter but comparable to the CFP2 in size. However, it will support a wider interface such that the module will support up to four wavelengths, each at speeds up to 400 Gb/s, for a total line-side capacity of 1.6 Tb per module [22].

5.5.1 Direct Detection Opportunity for Silicon Photonics

The data center interconnect platforms outlined in Section 5.4 all use coherent optics, the de facto transmission scheme used to transmit 100 Gb and faster wavelengths over distances of hundreds and thousands of kilomoters.

But many connections linking equipment between data centers are less than 100 km apart. Microsoft, e.g. classifies its data centers into two categories: linking switch buildings that make up a large-scale data center spread across a campus, and linking buildings across a metropolitan area. Linking adjacent buildings requires a 2-km link typically, whereas links in a metro area must span up to 70 km.

As Brad Booth, principal architect for Microsoft's Azure Global Networking Services, explains, coherent devices have been designed for ultralong-haul transmission, with all kinds of extra features. Such coherent devices have a relatively high power consumption, and need to be housed in a separate platform. When Microsoft looked at coherent optics for the data center, it concluded such optics was extremely costly, says Booth.

Instead what Microsoft wanted was a pluggable optical module that fits into its switch equipment to link data centers over both reaches, a connection that was very low cost but very high bandwidth, says Booth.

Accordingly, Microsoft dismissed coherent optics, choosing a simpler direct detection scheme instead. Coherent detection allows for the recovery at the receiver of all the information associated with the transmitted signal such as its phase and amplitude. Digital signal processing algorithms can then use this signal information to counter the impairments introduced on the channel [23]. This is why coherent optics always has an associated DSP-ASIC chip.

Direct detection has been the traditional scheme used for dense wavelength-division multiplexing optical transmission at 2.5, 10, and 40 Gb. With direct detection, the output of the photodetector at the receiver is an electrical current that is proportional to the squared measure (the complex electric field) of the received optical signal. What is important here is the inherent squaring introduced by the photo-detector; the squaring destroys signal information that is not available at the receiver but is available using coherent detection.

This information is important and explains why 100-Gb optical transport based on coherent optical transmission achieves a longer reach than 10-Gb direct detection despite the higher speed. But direct detection does the job for data center interconnect, where the link distances in question are 100 km or shorter.

Microsoft is working with chip company Inphi to develop a 100-Gb QSFP28 pluggable direct-detect module. Note, the QSFP28 module is mainly used for short-reach optics within the data center, not line-side optics for dense wavelength-division multiplexing optical transport.

The QSFP28 uses two wavelengths at 25 GBd/s combined with a modulation scheme (4-level pulse amplitude modulation, also known as PAM4) that encodes two bits on each signal duration or symbol. The result is 50 Gb per wavelength or 100 Gb overall. Inphi is using silicon photonics to implement what it calls the ColorZ QSFP28 module. Using the ColorZ will enable up to 4 Tb of capacity on a single fiber over a reach of up to 80 km. This design is referred to by Microsoft as Madison Phase 1.0.

Microsoft is in discussion with several vendors about developing a second-generation design, Madison 1.5, that will achieve more capacity in the C band by using 100 Gb over a single wavelength. Using multiple modules, capacity will be extended from 6.4 to 7.2 Tb over the C band.

A third design, Madison 2.0, will be a "coherent-lite" design, according to Booth, achieving speeds above the 100 Gb achieved using direct detection and PAM4. It will use 400-Gb lightpaths and achieve a total capacity of 38 Tb over the C band. The coherent optics will not fit inside a QSFP28 pluggable but will be implemented using on-board optics.

Microsoft is also leading an industry initiative known as the Consortium of On-Board Optics, or COBO, to develop standardized on-board optics. Such designs bring optics closer to a card's chips and increase the interface density of platforms—just what is needed for linking data centers. The COBO module would be placed next to the coherent-lite DSP-ASIC, or potentially the optics and the coherent chip could be built together [24,25].

Microsoft's Madison initiative is an example of an Internet content provider spurring an initiative that is leading to novel optical designs. The initiative also highlights how a semiconductor company—Inphi—can suddenly become a silicon photonics player. Lastly, it shows how the optical industry is delivering capacity in ever smaller form factors to drive down cost.

The Inphi product is one of many expected direct-detection module solutions.

Silicon photonics start-up Ranovus has announced a 200- Gb/s interface in a CFP2 form factor that will support links up to 130 km. The design uses four wavelengths each at 50 Gb/s using 25-GBd optics and PAM4 modulation. Up to 96 50-Gb channels can be fitted in the C band to achieve a total transmission bandwidth of 4.8 Tb [26].

5.6 TACKLING CONTINUAL TRAFFIC GROWTH

We have highlighted the importance of increasing line-side capacity for data center interconnect applications, but the issue of making best use of fiber capacity is also central for the telcos. This section looks at the approaches being considered to cope with continual traffic growth, both by increasing the capacity available within existing fiber networks and by developing new fiber optics.

The starting assumption is that the service providers, telcos and Internet players alike, will always seek to make best use of the equipment and fiber assets they already own; service providers only spend money when they have to.

5.6.1 Flexible Grid

One approach telcos are adopting is to move to a "flexible" grid. Operators traditionally have used fixed-sized channels across the fiber's C band spectrum, inside which sit the wavelengths or lightpaths that carry the data being transmitted (see Fig. 5.7). For dense wavelength-division multiplexing, these fixed channels are typically 50 GHz wide, allowing 96 wavelengths to fit across the C band. By having flexibility to position where these channel boundaries reside, 37.5-GHz wide channels can be used, e.g., allowing more wavelengths to be squeezed across the fiber's spectrum.

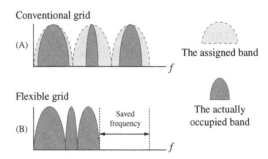

Figure 5.7 (A) Conventional and (B) flexible grid illustrated. Japanese Journal of Applied Physics.

We have already encountered this with Ciena's Waveserver data center interconnect product that supports 88 wavelengths for a total capacity of 17.6 Tb. Ciena achieves up to 25 Tb of capacity by using a flexible grid (as discussed in Section 5.4.2).

A major simulation study by telecom operator Telefonica showed that moving to a flexigrid network and getting rid of its rigid 50-GHz wide channels would delay the need for its network in Spain to be upgraded. Instead of an upgrade in 2019, introducing flexible grid ROADMs would support traffic growth till 2024 [27].

The UK telco BT conducted a trial with Chinese equipment vendor Huawei and showed that a 200-Gb lightpath can fit in a 33.3-GHz wide channel. BT expects that the C band can accommodate as much as 30 Tb but believes that this is close to the limit [28].

5.6.2 Flexible-Rate Transponders

One research development is a flexible-rate transponder that can adapt the modulation scheme used and hence the data carried over lightpaths depending on the communication channel. This concept, known as the sliceable bit-rate variable transponder, is still at an early stage [29]. The transponder would generate a very high-capacity transmission line, 10 Tb or greater. It would generate superchannels—optical channels made up of multiple wavelengths—that would be provisioned on demand.

The large multiterabit superchannel would be segmented within the network using flexible grid ROADMs that would direct parts of the superchannel to different destinations. Such a sliceable transponder promises several benefits:

- The multiterabit slice could be repartitioned based on demand. This would occur occasionally rather than in a highly dynamic way, adding extra capacity between destinations when needed. Accordingly, the sliced multiterabit superchannel would end up going to fewer destinations over time as a result of continual traffic growth.
- The sliceable transponder promises cost reduction through greater component integration. This is where indium phosphide and silicon photonics would come in, integrating multiple optical modulators on one large transponder, e.g., and even the lasers and receivers. Such an integrated transponder would clearly benefit data center interconnect platform density.

The sliceable transponder, should it become adopted, would only be deployed after 2020 at the earliest [29]. Much development work would be needed first to integrate 10 Tb into a transponder.

5.6.3 Using More Fiber Bands
After that, additional spectral bands in the fiber will need to be used. Using the L band is the next obvious step to increase the amount of data that can be sent across a fiber. Recall that Nokia and ADVA have already taken this approach for their platforms, supporting data transmission in the C and L bands. Telcos foresee using spectrum even wider than the C and L bands combined. This approach could extend the capacity of a single-mode fiber from 30 to 100 Tb (Table 5.1).

But once the nonlinear Shannon limit is reached in a transmission band, the only option is to light up a new fiber or band. Lighting a new fiber is fine if you are a telco that owns plenty of spare fiber, but it does not lead to greater efficiencies and the transport cost-per-bit rises, as detailed in Section 5.3. Optical research labs are also grappling with this issue and are already thinking about what can be done.

5.6.4 Space-Division Multiplexing: The Next Big Thing?
Space-division multiplexing promises to boost the capacity of a fiber by a factor of between 10 and 100 using parallel transmission paths. With space-division multiplexing, signals are physically

Table 5.1 Silica Optical Fiber Transmission Bands					
Name	O Band	E Band	S Band	C Band	L Band
Wavelength range (nm)	1260–1360	1360–1460	1460–1530	1530–1565	1565–1625
Note	Original band	Water peak band; loss reduced to below 0.5 dB/km		Erbium amplifier	Specially designed erbium amplifier

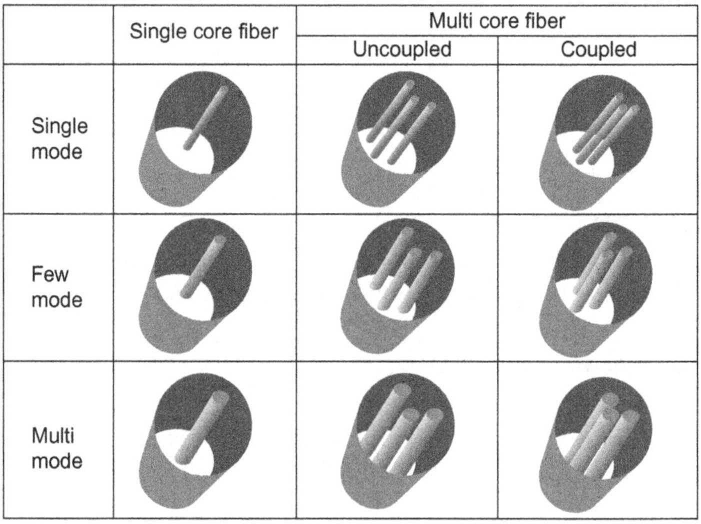

Figure 5.8 Conventional single-mode fiber and multicore fiber.[a]

aligned to the fiber and launched along these multiple paths (see Fig. 5.8).

The simplest way to create parallel paths is to bundle several standard fibers together. Note that this is different from simply adding and lighting a new fiber. The spatial-multiplexing equipment would use all the fibers, whereas lighting a new fiber would require a new optical transport system to be added at each end.

Alternatively, new types of fiber could be used that support multiple paths (cores) and even multiple light transmissions (optical modes) down each path. But this requires a new fiber type to be developed and installed, a hugely expensive endeavor.

[a]We would like to thank Susan Natacha González for her contribution to this figure.

Nokia Bell Labs claimed an industry first in late 2015 when it demonstrated a 60-km span of what it calls coupled 3-core fiber [30]. For the demo, Bell Labs generated twelve 2.5-Gb signals that were split down three paths, each path carrying in effect 10 Gb of data.

As suggested by the name—coupled 3-core fiber—the three signals in these cores interact strongly with each other. Digital signal processing is needed to restore the multiple transmissions at the receiver. To do this, Bell Labs employs a technique called multiple input, multiple output or MIMO that is already used for other forms of communications, such as cellular and digital subscriber line broadband. Bell Labs' industry-first achievement with this demo was being able to do MIMO processing for optical in real time. Until now, such high data rates required a brief transmission with the restoration at the receiver being done offline due to the intensive computation involved.

In effect, the transmission rotates the signals arbitrarily, while using MIMO at the receiver rotates them back. The signals are garbled up due to the rotation, and undoing the rotation is called MIMO, explains Bell Labs' Peter Winzer.

Will operators adopt such advanced technology and deploy a new style of fiber? Spatial multiplexing is a topic that to date has largely been dismissed as an extravagant research activity rather than a technology that promises to solve, or at least postpone, a looming capacity crunch.

Even if it does get deployed, it will not be for another decade and likely longer, and it will be expensive. Yet besides adding spectral bands and then more fiber, the industry is not offering an alternative to keep scaling capacity.

Telcos are not about to start replacing their huge worldwide fiber cable investments with novel fiber that has yet to even be specified. But simply lighting a new single-mode fiber and adding optical transport equipment at each end will not reduce the cost-per-bit, as discussed. Only by further integrating spatial multiplexing systems will cost come down by sharing equipment across multiple parallel transmission paths.

This is what Bell Labs is tackling. Winzer says that to get the cost-per-bit down, new levels of integration will be needed. Integration will happen first across the transponders and amplifiers; fiber will come last, he says.

This is just the sort of opportunity where silicon photonics can lead. This development is a decade out and will require novel designs, and 10 years is plenty of time for silicon photonics to become more commercially mature.

5.7 PULLING IT ALL TOGETHER

Network capacity is becoming a key challenge facing service providers due to the continual growth in traffic. Fiber capacity, while vast, is finite and is now being exhausted. Transport cost is set to rise as new network investments will be required for common equipment shared by all the wavelengths, not just new equipment at the fiber end points. Capacity exhaust is the main issue, not the typical system design issues of power, cost, and size reduction, although all these play a role too.

The strategies available to the telcos and Internet content providers are to exploit the full capacity of the existing installed fiber and only then pay for the network upgrade. Then the story becomes cost optimization.

This will likely be done in two stages. First, cost-reducing the end equipment based on power consumption and size will require continual integration to bring more capacity into smaller and smaller form factors. This means working to fit the current capacity of a fiber into a single shelf and ultimately into a module and a chip. Second, new techniques to expand the transmission capacity of current fiber will be needed. Spatial multiplexing, a promising technique but still far from commercialization, will require unprecedented degrees of integration to feed light down a fiber's multiple paths and modes.

Silicon photonics has a role in all these developments. As has been highlighted, the Internet content providers have become important technology drivers. But cost is key. If silicon photonics can enable cheaper integrated solutions, it will finally leapfrog indium phosphide for what many consider the endgame of line-side transmission.

Key Takeaways

- The Internet content providers are shaking up the traditional telecom market. Unlike the telcos, these newer players are experiencing sharp revenue growth.
- The Internet content providers are also experiencing far faster traffic growth than the telcos. The telcos' core network traffic is growing annually at between 20% and 30%, while for the Internet content providers it is more like 80−100%.
- The requirements of the Internet content providers have given rise to a new category of optical transport platform, known as data center interconnect. These platforms condense as much line-side capacity into as small a space as possible.
- The industry should prepare for the cost per transported bit to rise once the C band fills up.
- Optical designers need to develop equipment that drives down the cost of transporting bits. This requires greater performance efficiencies, made all the more challenging as systems use more complex signaling techniques to cope with the continual growth in traffic.
- To advance capacity further, other fiber spectral bands besides the C band will be needed. After that, the next option is moving to new fiber, either conventionally by lighting fresh fiber or using spatial multiplexing and parallel transmission paths.
- Space-division multiplexing requires a new style of optical design, with integration taking place across these parallel transmission paths. To make such systems cost-effective and commercial will require significant investment, development work, and integration.
- Silicon photonics is already being deployed for line-side optics. It is also being used for linking data centers: in coherent optics with data center interconnect and direct detection modules with a range of up to 130 km.
- Silicon photonics has yet to demonstrate a telling advantage when compared to indium phosphide. And given that line-side optics volumes will be relatively modest, silicon photonics will find it hard to differentiate itself, at least in the coming years.
- System vendors are better positioned than the optical component vendors to drive forward silicon photonics development for line-side optics. This is also true in the data center, as will be described in Chapter 7, Data Center Architectures and Opportunities for Silicon Photonics.
- We expect silicon photonics to play an increasingly important role, but we do not yet see it displacing indium phosphide. The Layer 4 telecom network trends will only benefit the technology going forward, and silicon photonics could gain an edge longer term.

REFERENCES

[1] 60-second interview with Infonetics' Andrew Schmitt. Gazettabyte, < http://www.gazetta-byte.com/home/2015/2/26/60-second-interview-with-infonetics-andrew-schmitt.html > ; February 26, 2015.

[2] Silicon carves out its niche as an optical material. Fibre Systems Europe; March–April 2009. p. 18–21.

[3] Alexander Graham Bell demonstrates AT&T's transcontinental telephone line. The New York Times, < http://learning.blogs.nytimes.com/2012/01/25/jan-25-1915-alexander-graham-bell-demonstrates-atts-transcontinental-telephone-line/?_r = 0 > .

[4] AT&T completes acquisition of DIRECTV, press release, < http://about.att.com/story/att_-completes_acquisition_of_directv.html > ; July 24, 2015

[5] NTT confirms it will buy e-shelter. DatacenterDynamics, < http://www.datacenterdynamics.com/content-tracks/design-build/ntt-confirms-it-will-buy-e-shelter/93481.fullarticle > ; March 3, 2015.

[6] Verizon buys Yahoo: does it add up? The Economist; July 30, 2016. p. 47.

[7] SoftBank to buy ARM holdings for $32 billion. Wall Street Journal, < http://www.wsj.com/articles/softbank-agrees-to-buy-arm-holdings-for-more-than-32-billion-1468808434 > ; July 18, 2016.

[8] SDM and MIMO: an interview with Bell Labs. Gazettabyte, < http://www.gazettabyte.com/home/2015/12/18/sdm-and-mimo-an-interview-with-bell-labs.html > ; December 18, 2015.

[9] Nokia's PSE-2s delivers 400 gigabit on a wavelength. Gazettabyte, < http://www.gazetta-byte.com/home/2016/6/8/nokias-pse-2s-delivers-400-gigabit-on-a-wavelength.html > ; June 8, 2016.

[10] Nokia ushers in 100G transport services era with new programmable silicon chipset and next-generation optical networking systems, Nokia press release, < http://company.nokia.com/en/news/press-releases/2016/03/16/nokia-ushers-in-100g-transport-services-era-with-new-programmable-silicon-chipset-and-next-generation-optical-networking-systems > ; March 16, 2016.

[11] Shannon CE. A mathematical theory of communication. Bell Syst Tech J 1948;27:379–423 623–656

[12] Essiambre R, et al. Capacity limits of optical fiber networks. J Lightwave Technol 2010;28 (4):662–701.

[13] Ovum forecast. Optical networks forecast spreadsheet: 2014–19, TE0006-000982; January 2015.

[14] Infinera targets the metro cloud. Gazettabyte, < http://www.gazettabyte.com/home/2014/11/6/infinera-targets-the-metro-cloud.html > ; November 6, 2014.

[15] Ciena's stackable platform for data centre interconnect. Gazettabyte, < http://www.gazetta-byte.com/home/2015/5/31/cienas-stackable-platform-for-data-centre-interconnect.html > ; May 31, 2015.

[16] ADVA's 100 terabit data centre interconnect platform. Gazettabyte, < http://www.gazetta-byte.com/home/2015/6/9/advas-100-terabit-data-centre-interconnect-platform.html > ; June 9, 2015.

[17] Coriant's 134 terabit data centre interconnect platform. Gazettabyte, < http://www.gazetta-byte.com/home/2015/12/17/coriants-134-terabit-data-centre-interconnect-platform.html > ; December 17, 2015.

[18] Cisco NCS 1002 data sheet, < http://www.cisco.com/c/en/us/products/collateral/optical-net-working/network-convergence-system-1000-series/datasheet-c78-733699.html > .

[19] Infinera introduces Cloud Xpress 2 with infinite capacity engine; Extends Leadership in Data Center Interconnect, press release, < http://investors.infinera.com/new-releases/press-release-details/2016/Infinera-Introduces-Cloud-Xpress-2-With-Infinite-Capacity-Engine-Extends-Leadership-in-Data-Center-Interconnect/default.aspx > ; September 21, 2016.

[20] Infinera goes multi-terabit with its latest photonic IC. Gazettabyte, < http://www.gazettabyte.com/home/2016/3/30/infinera-goes-multi-terabit-with-its-latest-photonic-ic.html > ; March 30, 2016.

[21] Next-generation coherent adds sub-carriers to capabilities. Gazettabyte, < http://www.gazettabyte.com/home/2016/1/24/next-generation-coherent-adds-sub-carriers-to-capabilities.html > ; January 24, 2016.

[22] OIF starts work on a terabit-plus CFP8-ACO module. Gazettabyte, < http://www.gazettabyte.com/home/2016/7/24/oif-starts-work-on-a-terabit-plus-cfp8-aco-module.html > ; July 24, 2016.

[23] Roberts K, et al. Performance of dual polarization QPSK for optical transport. J Lightwave Technol 2009;27(16):3546−59.

[24] COBO website, < http://cobo.azurewebsites.net/ > .

[25] COBO looks inside and beyond the data centre. Gazettabyte, < http://www.gazettabyte.com/home/2015/11/25/cobo-looks-inside-and-beyond-the-data-centre.html > ; November 25, 2015.

[26] Ranovus shows 200 gigabit direct detection at ECOC. Gazettabyte, < http://www.gazettabyte.com/home/2016/9/20/ranovus-shows-200-gigabit-direct-detection-at-ecoc.html > ; September 20, 2016.

[27] Fernandez-Palacios JP, et al. Elastic optical networking: an operators perspective, ECOC 2014, < http://www.vlopezalvarez.com/Profesional/Publications/Conferences/2014_ECOC_6.pdf > .

[28] BT & Huawei achieve real-world super channel speed of 3Tbps, < http://www.huawei.com/uk/about-huawei/newsroom/press-release/hw-416171.htm; October 8, 2015.

[29] BT makes plans for continued traffic growth in its core. Gazettabyte, < http://www.gazettabyte.com/home/2016/1/19/bt-makes-plans-for-continued-traffic-growth-in-its-core.html > ; January 19, 2016.

[30] Randel S, et al. First real-time coherent MIMO-DSP for six coupled mode transmission. In: IEEE photonics conference (IPC), October 4−8, 2015; 2015.

CHAPTER 6

The Data Center: A Central Cog in the Digital Economy

We are no longer in a linear one-dimensional industry, we are in a dynamic multi-faceted one. The fluidity at the moment is all around the data center, open networking, and how the unconventional network operators are influencing the market.

Brandon Collings, CTO of Lumentum

6.1 INTERNET CONTENT PROVIDERS ARE DRIVING THE NEW ECONOMY

The combination of the Internet, social media, and mobile devices has created an economy that offers entrepreneurs new ways to make money. Fig. 6.1 is a snapshot of the broad spectrum of Internet applications and the pervasiveness of their usage [1]. Internet content providers such as Google, Facebook, Amazon, Microsoft, and Apple are leading companies in this new economy, competing fiercely for the business of consumers and enterprises.

Data centers play a fundamental role in generating revenues for the Internet content providers. They are the ultimate modern-day factories that take in raw material (information in the form of digital data) and use labor (server-based computing and information processing) to manufacture products (services and data) that are distributed in the marketplace (via the Internet).

Data centers, packed with servers, must support annual double-digit traffic growth to ensure the delivery of services efficiently to users and businesses. Moreover, the Internet content providers' ability to capture market share depends on their capability to fine-tune their data centers to maximize efficiencies. For now, revenue growth of the Internet content providers shows no sign of abating.

Silicon Photonics. DOI: http://dx.doi.org/10.1016/B978-0-12-802975-6.00006-5
Copyright © 2017 Daryl Inniss and Roy Rubenstein. Published by Elsevier Inc. All rights reserved.

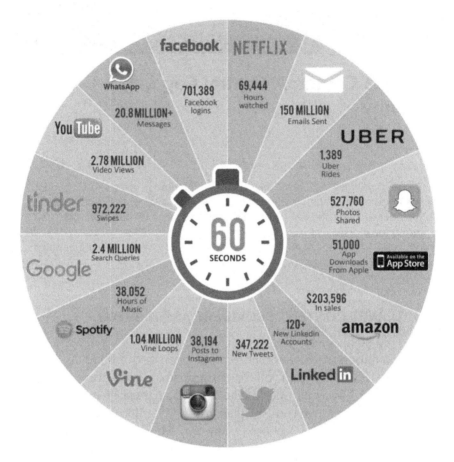

Figure 6.1 What happens in a 2016 Internet minute? Courtesy of Excelacom.

Internet content providers are disrupting several markets. After its first 20 years, spending on Internet advertising in the United States outpaced radio and television advertising during their first 20 years (normalized to 2015 revenues). The Internet accounted for $60 billion in advertising revenues in 2015 in the United States alone. Internet content providers Google and Facebook accounted for over three quarters of these revenues, and their share is growing [2].

E-commerce is another market segment where the Internet is having a huge impact, with growth forecast to continue. Internet sales accounted for 12% of all US retail—$340 billion—in 2015 [2].

The ability of the Internet content providers to reliably and cost-effectively deliver services to the user is key to this financial success, and data centers and associated networking are the enablers. For this web economy to continue to grow, it is important to offer new services to a broader community of customers while ensuring their experience is a good one.

The data center and its growing network bandwidth requirements offer silicon photonics the best near-term market opportunity. Silicon photonics promises to deliver cheaper, more compact, lower power high-capacity optical circuits—core infrastructure enablers needed to continue to fuel this new economy.

This chapter highlights the digital economy opportunities and the consequences and challenges resulting from the growth in the number and size of data centers. Silicon photonics is a technology suited to enable data centers to continue scaling, as highlighted in Chapter 5, Metro and Long-Haul Network Growth Demands Exponential Progress, and expanded upon in Chapter 7, Data Center Architectures and Opportunities for Silicon Photonics.

6.2 CLOUD COMPUTING: ANOTHER GROWTH MARKET

Cloud computing—offering IT and telecom services over the network—is another growth business opportunity for the Internet content providers. Cloud computing provides services for businesses, from small enterprises to multinationals, and for consumers. Cloud computing services can be offered over the same data center infrastructure used for existing services such as Internet searches for Google or social networking for Facebook. The infrastructure can be adapted to make available computing, applications, and storage resources over the Internet, creating an additional revenue stream.

Amazon Web Services offers a glimpse into this revenue potential. Amazon Web Services is the recognized leader in the cloud-based services market [3], and its 2013−15 annual revenues are illustrated in Fig. 6.2. Annual sales nearly tripled in 2 years, and if the trend continues revenues will exceed $10 billion in 2016.

Amazon Web Services has been building data centers worldwide to support its cloud business. The company operates 38 "Availability Zones" in 14 geographical regions, where each zone consists of one or more data centers [4]. Amazon plans to introduce nine more zones in

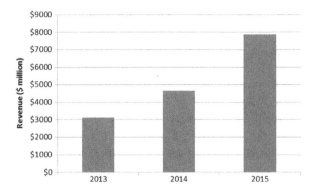

Figure 6.2 Amazon Web Services' annual revenues: 2013–15. Courtesy of Amazon.

four more regions throughout 2017, and it continues to expand its business in Europe and Asia.

In effect, Amazon Web Services is developing a scale of operation needed to deliver secure and low-latency cloud services. Amazon understands what is needed to support customer demand, with its data centers playing a key role.

6.2.1 Microsoft, Google, and Others Seek to Grow Their Cloud Businesses

Amazon Web Services' nearest competitor in the cloud service business, Microsoft, is trying to grab more market share for its Azure cloud service. Financial analysts estimate Microsoft's Azure revenues were $500–$700 million in 2015, a tenth of Amazon Web Services [5]. Microsoft, famous for its personal computer operating systems and software, boasts 1.5 million servers housed in 100-plus data centers globally [6,7].

Google is also seeking to win cloud computing business with its Google Cloud Platform, promising customers use of the same infrastructure that powers Google search and YouTube services. Google has captured from Amazon some of Apple's cloud business, a significant feather in its cap [8].

The Internet content providers also use each other's data centers to buttress their own offerings. Apple, e.g., was already a customer of Amazon Web Services and Microsoft Azure before signing on with Google [9]. Apple is rumored to be building data centers to expand its cloud offerings and reduce its dependency on sourced support.

The expansion of data centers to grow cloud businesses is not confined to the North American Internet content providers. Tencent, an important Chinese Internet content provider, announced that it is spending $1.6 billion to build data centers in China, Hong Kong, and North America to expand its cloud computing business [10].

Moreover, Tencent is collaborating with Intel on its Rack Scale Architecture—which disaggregates the servers' computational, storage, memory, and network functions—to demonstrate the technology's performance. While Tencent continues its assessment of the server architecture, it has reported better resource utilization and lower total cost of ownership [11].

6.2.2 Telcos Offer a Mixed Story Regarding Cloud Computing

Telcos are also active in growing their cloud service businesses and winning consumer and enterprise customers.

The data center strategy of the telcos has changed, however. Mergers and acquisitions ruled the day in 2011 for North American carriers seeking to expand their cloud offerings and credentials. Verizon's acquisition of Terremark for $1.4 billion [12] and CenturyLink's merger with Savvis for $2.5 billion were marquee transactions then [13].

More recently, telecom service providers have investigated selling these assets in what appears to be a reversal of their initial strategy. For example, AT&T's data centers were rumored to be for sale in 2015, and eventually the operator struck a deal with IBM to take over managing its hosting service [14]. IBM is acquiring the equipment used to host the services. Both CenturyLink [15] and Verizon [16] are also considering selling their data centers.

One reason for this strategy switch is that the North American telcos have significant debt, a consequence of purchasing radio spectrum to expand their mobile services [17]. The telcos recognize they need not own and operate data centers to continue offering colocation and cloud services (colocation refers to offering third parties a presence in a data center). The telcos have also sold other assets to service their debt: Verizon, e.g., sold part of its wireline business to Frontier Communications for $8.6 billion [18], and AT&T and Verizon are selling their cell towers and leasing them back to support their mobile networks [17].

While using hosting services to support cloud services is an approach taken by some telcos, this is not a universal rule. Japan's NTT, e.g., has acquired data centers globally. NTT sees its cloud business as central to its plan to grow its overseas business and as a key pillar of its overall strategy [19].

Telcos have central offices that can be rearchitected as data centers. These are not large-scale data center venues but are valuable resources to have. Central offices are used by telcos to house generations of equipment developed for their telecom services. Telecom carriers can own thousands of central offices—AT&T alone has between 4000 and 5000 [20]—that serve tens to hundreds of thousands of customers and enterprises. But they are not tooled to address today's emerging opportunities based on software-defined networking and network functions virtualization, which are being used to support telecom service implementation on generic servers. In effect, telcos have seen the merits of running operations as the Internet content providers do, and they are looking to do the same for their own services.

AT&T is driving a high-profile program to transform its central offices to data centers. The program, known as Central Office Re-Architected Datacenter or CORD [21], is focused on the access network, delivering services on a passive optical network, e.g., to consumers and enterprises alike. And since these data centers are distributed and located at the edge of the telcos' network, they provide an excellent opportunity to offer services close to customers.

6.2.3 Traditional Enterprises Use a Hybrid Cloud Model

To manage costs and deliver services to many regions, enterprises use data centers from providers such as Equinix and Digital Realty that allow connection to multiple telecom service providers. Enterprises prefer to use a hybrid cloud—i.e., a combination of private or on-premises and public or third party–based cloud services. This model allows them to have distributed data centers without the cost of building and managing them in all regions.

An announced deal between Shaw Communications and Microsoft in 2016 exemplifies such a relationship [22]. Customers can colocate servers in the Shaw data center while having the option of a public cloud service such as Azure.

6.3 THE EXPANSIVE BUILD-OUT OF DATA CENTERS

Internet content providers, telcos, and enterprises are all building data centers as well as seeking extra capacity through third parties as they expand their scale and offerings. Fig. 6.3 shows Facebook's two Prineville, Oregon, data centers, and a third building, 40% larger than the first two, has been announced [23]. These buildings are distributed, and the growing bandwidth requirements they demand benefit silicon photonics, as outlined in Chapter 5, Metro and Long-Haul Network Growth Demands Exponential Progress.

6.3.1 Demand for Video Adds to Bandwidth Pressures

One trend helping to drive data center expansion is the growth of video. Video is a well-known network bandwidth hog, with Netflix's streaming video service being the biggest culprit followed by Google's YouTube [24]. Leading social media providers are also getting in on the act, introducing and extending services based on user-generated video.

The growth in daily video views on Facebook is shown in Fig. 6.4. It increased two and a half times during the period shown, from 3.2 billion views of at least 3 seconds duration at the end of 2014 to 8.0 billion views in the third quarter of 2015. And this does not include Facebook Live, a service launched in April 2016 that allows users to share live video and is driving further usage.

Figure 6.3 Facebook's two 340,000-square foot Prineville, Oregon, data centers. © Copyright Facebook.

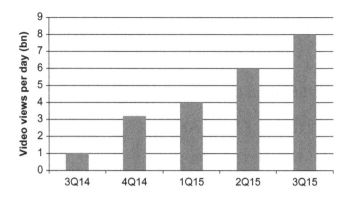

Figure 6.4 Average video views per day on Facebook, 3Q14–3Q15. From KPCB Internet Trends 2016.

Meanwhile, Snapchat (since renamed Snap), another social media site, reported 10 billion video views a day in the first quarter of 2016 [25].

Facebook and Snap users are generating video that is streamed over the network, requiring service providers to support video-level bandwidth with storage and networking while maintaining the ability to rebroadcast clips on demand. Such traffic growth will increasingly stretch the networks of both the Internet content providers and the telcos. These services are in their infancy and will have a significant impact on networks and data centers.

Not surprisingly given these trends, Internet content providers are spending billions of dollars on building new data centers. In many cases they are building campuses where hundreds of thousands of servers are spread across several buildings. Microsoft, e.g., already has two data centers in San Antonio, Texas, and is said to be spending $1 billion to build a third, covering multiple buildings for a total floor space of 1.2–1.3 million square feet [26]. Breaking ground in 2016, the target completion date is between 2021 and 2023. And Apple, Google, Facebook, and Amazon are all building large-scale data centers to support their cloud services.

6.3.2 Facebook's Data Center Builds: A Case Study

Facebook operates four campuses in the United States, each made up of multiple data center buildings, each totaling over 100,000 square feet. And Facebook is bringing data centers online around the world. Fig. 6.5 shows Facebook's operational data centers and those

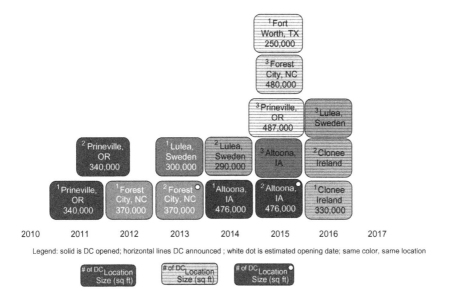

Figure 6.5 Facebook's data centers: in operation and planned.

announced, along with data center locations and approximate size (where data is available).

Building multiple data centers in one location is a clear trend, as illustrated in Fig. 6.5, and is a practice being followed by many Internet content providers. Typically, campus building connections are up to 2 km, distances that can be supported by 100 Gb/s silicon photonic transceivers that are commercially available today. And, as discussed in Chapter 5, Metro and Long-Haul Network Growth Demands Exponential Progress, new direct detection dense wavelength-division multiplexed modules are being developed to offer high-capacity links between buildings.

Fig. 6.6 highlights the increasing square footage of data centers. The data in Fig. 6.5 was used to construct Fig. 6.6 with the assumption that announced data centers will be operational 2 years after construction starts.

The increasing size of data centers comprising several buildings comes at a cost. Rising power consumption is one sensitive issue because it impacts the cost of doing business: operational cost.

Like other Internet content providers, Facebook is sensitive to its data centers' environmental impact and energy demands. It has chosen

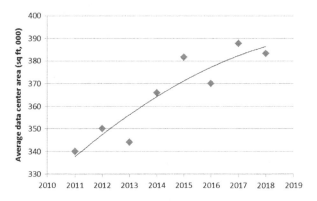

Figure 6.6 Facebook's average data center square footage, 2011–18.

locations like Lulea in Sweden where it can use the cold climate's air to cool its servers. Wind power and evaporative cooling are being used in other locations, and Facebook is developing facilities that do not require traditional air conditioning.

Facebook also partners with local power authorities and is building renewable energy nearby for its data centers to drive down their operational cost and environmental impact. The issue of power consumption is now discussed.

6.4 ENERGY CONSUMPTION POSES THE GREATEST DATA CENTER CHALLENGE

Large-scale data centers consume vast amounts of power. Facebook's three data centers in Prineville, Oregon, use some 70 MW [27], about two-thirds the power used by all the homes in the rest of the Oregon county where the data centers are located [28]. For reference, 100 MW can power 80,000 US homes [29].

Many Internet content providers use renewable energy including solar, wind, and natural cooling. According to environmental group Greenpeace, Google is the leading Internet player in purchasing renewable energy, with an estimated 35% of Google's operations using clean energy [27].

In addition to using renewable energy and natural means of cooling, data center operators also seek greater efficiencies from their powered equipment. Server usage has been increased through software

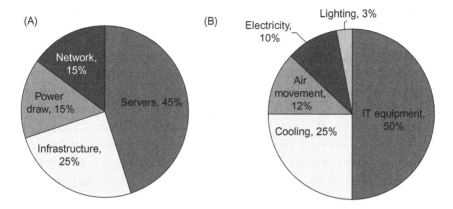

Figure 6.7 Typical data center costs and energy consumption breakdowns. (A) Data center costs and (B) data center energy consumption. (A) Data from Microsoft (B) Data from Computerworld from EYP Mission Critical Facilities Inc, New York.

innovations and network sharing. Data center managers also replace equipment on much shorter timescales, from every 5–7 years previously to 2–3 years today. New equipment is more energy-efficient while delivering higher-bandwidth performance [30].

Energy remains a key issue when building large-scale data centers, but despite technological advances, the rate at which the power consumption requirements of the data centers is growing is a major concern. Electricity consumption from data centers and their networking was about 700 billion kWh in 2012 and is expected to increase to over 1200 billion kWh by 2017, according to Greenpeace [27].

Within the data center, the server is the most costly item (Fig. 6.7A), and it is the most power-demanding equipment class, accounting for nearly half of the power consumption as illustrated in Fig. 6.7B; IT equipment is dominated by the server. The second largest power-hungry item is the heating, ventilation, and air conditioning (HVAC) needed to keep the room temperature consistent so the IT equipment such as servers and the switches that connect them operate correctly.

As bandwidth demand continues to increase, and as Moore's law ends, data center operators need to develop power-efficient strategies and reduce overall data center size while driving up the bandwidth that can be supported.

6.5 SILICON PHOTONICS CAN ADDRESS DATA CENTER CHALLENGES

Notwithstanding improvements in design and operation, Greenpeace finds data center power consumption the fastest growing among three important segments: end-user devices, networking, and the data center. More bandwidth is needed to support the growth in video and the increasing number of consumers using broadband. What is needed are ways to deliver services using more compact, power-efficient systems. This is an opportunity for systems innovation. How can silicon photonics help? This is the topic of Chapter 7, Data Center Architectures and Opportunities for Silicon Photonics.

Key Takeaways

- Internet content providers are transforming industries with Internet-based service platforms. Tens of billions of dollars of revenues are being made in advertising and cloud computing, with revenue growth set to continue.
- The data center is the central cog that fuels the Internet content providers' businesses. These companies are spending billions on data centers to increase capacity and support continued service growth.
- Telecom operators and enterprises are interested in cloud-based services but are split on the merits of owning data centers. Telecom operators that previously acquired data centers are rethinking whether owning them is core to their businesses. Enterprises also own data centers but usually rely on a hybrid cloud model to deliver their services.
- New video services are being introduced, further driving requirements and fueling demand for data centers. The Internet content providers are responding by building more and larger data centers.
- Energy demand is a core challenge when it comes to scaling data centers. Numerous steps are being taken to reduce power consumption. These include renewable energy, building at locations where natural cooling is used, and upgrading equipment frequently.
- Servers are the largest consumers of energy in the data center, with heating, venting, and cooling being the next highest.
- Data center capacity must continue to scale to support the enormous capacity demand that is coming. Innovation is needed, and silicon photonics is a technology that promises to rise to the challenge as described in the following chapter.

REFERENCES

[1] 2016 Update: what happens in one Internet minute?, < http://www.excelacom.com/resources/blog/2016-update-what-happens-in-one-internet-minute > ; February 29, 2016.

[2] KPBC Internet trends 2016, < http://www.kpcb.com/internet-trends > , slides 54 and 44.

[3] Amazon financial reports, < http://www.recode.net/2016/4/28/11586526/aws-cloud-revenue-growth > .

[4] AWS global infrastructure, < https://aws.amazon.com/about-aws/global-infrastructure/ > .

[5] Largest cloud provider in 2016—AWS vs Azure, < http://stratusly.com/iaas-market-share-in-2016-aws-vs-azure/ > ; December 9, 2015.

[6] Microsoft global datacenters 2016, < https://www.microsoft.com/en-us/cloud-platform/global-datacenters#Fragment_Scenario1 > , Brad Booth interview.

[7] Azure regions, < https://azure.microsoft.com/en-us/regions/ > .

[8] Cloud makes for strange bedfellows: Apple signs on with Google, cuts spending with AWS, < http://www.crn.com/news/cloud/300080062/cloud-makes-for-strange-bedfellows-apple-signs-on-with-google-cuts-spending-with-aws.htm?itc = refresh > and Google gets Apple to jump aboard its cloud business (though it may not last), < http://recode.net/2016/03/16/google-gets-apple-to-jump-aboard-its-cloud-business-though-it-may-not-last/ > ; March 16, 2016.

[9] iOS Security iOS 9.3 or later, < http://www.apple.com/business/docs/iOS_Security_Guide.pdf > ; May 2016, p. 43.

[10] China's tencent to invest $1.57 billion in cloud computing over five years, < http://www.wsj.com/articles/chinas-tencent-to-invest-1-57-billion-in-cloud-computing-over-five-years-1442320553 > ; September 15, 2015.

[11] Tencent explores datacenter resource-pooling using Intel Rack Scale Architecture, < http://www.intel.com/content/dam/www/public/us/en/documents/white-papers/rsa-tencent-paper.pdf > ; 2015.

[12] Verizon completes Terremark acquisition. Verizon press release, < http://www.verizon.com/about/news/press-releases/verizon-completes-terremark-acquisition > ; April 11, 2011.

[13] CenturyLink to acquire Savvis for $40 per share in cash and stock transaction. CenturyLink press release, < http://news.centurylink.com/news/centurylink-to-acquire-savvis-for-40-per-share-in-cash-and-stock-transaction > ; April 26, 2011.

[14] AT&T and IBM expand their strategic relationship in managed application and managed hosting services. IBM press release, < https://www-03.ibm.com/press/us/en/pressrelease/48500.wss > ; December 17, 2015.

[15] CenturyLink continues expanding data centers it may sell. Data Center Knowledge, < http://www.datacenterknowledge.com/archives/2016/03/21/centurylink-continues-expanding-data-centers-it-may-sell/ > ; March 21, 2016.

[16] Verizon confirms it may sell data centers. Data Center Knowledge, < http://www.datacenterknowledge.com/archives/2016/01/21/verizon-confirms-may-sell-data-centers/ > ; January 21, 2016 and Verizon CFO says data center sale 'progressing,' Telco will make decision in next 3 months, < http://www.fiercetelecom.com/story/verizon-cfo-says-data-center-sale-progressing-telco-will-make-decision-next/2016-06-07 > ; June 7, 2016.

[17] Why AT&T and Verizon communications are selling over $12.6 billion worth of assets. The Motley Fool, < http://www.fool.com/investing/general/2015/02/15/why-att-and-verizon-communications-are-selling-ove.aspx > ; February 15, 2015.

[18] Verizon completes sale of landline assets in California, Florida and Texas to frontier communications. Verizon press release, < http://www.verizon.com/about/news/verizon-completes-sale-landline-assets-california-florida-and-texas-frontier-communications > ; April 1, 2016.

[19] e-shelter Deal makes NTT Europe's third-largest data center provider, < http://www.datacenterknowledge.com/archives/2015/03/03/e-shelter-deal-makes-ntt-third-in-europe-data-center-market/ > ; March 3, 2015; NTT Communications acquires 86.7% stake in e-shelter, Germany's top data-center operator. Data Center Knowledge. NTT Communications press release, < http://www.ntt.com/en/about-us/press-releases/news/article/2015/20150623.html > ; June 23, 2015; NTT Communications completes investment of $350 million in Ragingwire Data Centers to Acquire 80% Ownership Stake, < http://www.ragingwire.com/news/ragingwire-joins-ntt-family-of-companies > ; February 4, 2014; and NTT Expands Data Center, Cloud to U.S. With Acquisition, < http://www.pcworld.com/article/2058540/ntt-expands-data-center-cloud-to-us-with-acquisition.html > ; October 28, 2013.

[20] Central Office re-architected as a datacenter (CORD), < http://onrc.stanford.edu/protected %20files/PDF/ONRC-CORD-Larry.pdf > .

[21] CORD, < http://opencord.org/ > .

[22] Shaw to offer Microsoft Cloud out of viawest data centers. Data Center Knowledge, < http://www.datacenterknowledge.com/archives/2016/01/07/shaw-to-offer-microsoft-cloud-out-of-viawest-data-centers/ > ; January 7, 2016.

[23] Facebook greenlights third Prineville data center: it's Oregon's biggest. The Oregoneon, < http://www.oregonlive.com/silicon-forest/index.ssf/2015/09/facebook_greenlights_third_pri. html > ; August 14, 2015.

[24] Netflix Bandwidth Usage Climbs to Nearly 37% of Internet Traffic at Peak Hours. Variety, < http://variety.com/2015/digital/news/netflix-bandwidth-usage-internet-traffic-1201507187/ > ; May 28, 2015.

[25] Snapchat user 'stories' fuel 10 billion daily video views. Bloomberg, < http://www.bloomberg. com/news/articles/2016-04-28/snapchat-user-content-fuels-jump-to-10-billion-daily-video-views > ; April 28, 2016.

[26] Microsoft to build Huge Texas Data Center Campus. Data Center Knowledge, < http://www.datacenterknowledge.com/archives/2015/12/10/report-microsoft-to-build-huge-texas-data-center-campus/ > ; December 10, 2015 and Microsoft Buys Nearly 160 Acres in Far West San Antonio for Data Center Development. San Antonio Business J, < http://www. bizjournals.com/sanantonio/news/2015/12/10/exclusive-microsoft-buys-nearly-160-acres-in-far.html?ana = twt > ; December 10, 2015.

[27] Clicking clean: a guide to building the green Internet. Greeenpeace, < http://www.greenpeace. org/usa/wp-content/uploads/legacy/Global/usa/planet3/PDFs/2015ClickingClean.pdf > ;   May 2015, p. 43.

[28] Facebook discloses power use in Prineville and across the company, Wins Praise From Greenpeace. The Oregonian, < http://www.oregonlive.com/silicon-forest/index.ssf/2012/08/ facebook_discloses_power_use_i.html > ; August 1, 2012.

[29] The era of the 100 MW data center. Gigaom, < https://gigaom.com/2012/01/31/the-era-of-the-100-mw-data-center/ > ; January 31, 2012.

[30] Why cloud is good for the environment. Fujitsu, < http://www.fujitsu.com/ie/Images/wp-eiu-impact-of-cloud.pdf > ; p. 17.

Data Center Architectures and Opportunities for Silicon Photonics

Everything that has happened in the telecom network is now being replicated inside the data center. And then everything that is happening in the data center is going to be on the board, and then everything on the board is going to be in the package, and then everything in the package is going to be on the chip.

Lionel Kimerling

Data analytics is driving so much bandwidth and so much traffic within data centers, and it has such commercial value these days. I don't want to be too clichéd but it acts almost as the 'killer app' for driving photonics.

Keren Bergman

7.1 INTRODUCTION

In his book, *Thing Explainer*, Randall Munroe uses diagrams and labels to explain how complicated systems work [1]. A novelty of the book is that the author confines himself to a 1000-word vocabulary. For Munroe, the data center is a *computer building*, a server rack is a *holder*, and the server's processor is a *thinking box*.

We view the data center as the factory of the information age. The raw material is data which, when processed using energy and computation, is transformed into information. The factory's output may be the result of a complex data analytics algorithm or a service delivered to an enterprise or a consumer, e.g., the WolframAlpha application discussed in Chapter 2, Layers and the Evolution of Communications Networks.

As with all factories, improving efficiency and adding an assembly line benefit the factory owner's bottom line. The computing power, switching, and storage capacity needed to deliver the product, or

Silicon Photonics. DOI: http://dx.doi.org/10.1016/B978-0-12-802975-6.00007-7
Copyright © 2017 Daryl Inniss and Roy Rubenstein. Published by Elsevier Inc. All rights reserved.

service, is the equivalent of a factory assembly line, and the software that runs on the servers can be created and scaled dramatically in ways traditional manufacturing cannot match.

The data center also has few moving parts and is highly automated. Instead of rhythmic machinery, it has ventilation noise, blinking lights, and heat generated by rows of equipment. But the lack of an assembly line buzz should not be mistaken for inactivity—data centers are hugely complex, productive, and challenging environments.

The data centers continue to grow in size as they house ever more processors and racks, as discussed in Chapter 6, The Data Center: A Central Cog in the Digital Economy. The tasks the computing resources perform are also evolving. Such demands challenge the networking of servers. For data center managers, scaling operations cost-effectively remains a major headache.

The rising demand for video services is one development driving the need for more bandwidth. Fig. 7.1 shows the projected increase in cloud-based video and the growing appetite for viewing video by end users [2]. The bandwidth needed, and the uncertainty as to when and where video demand may originate, tax a data center's performance and its networking.

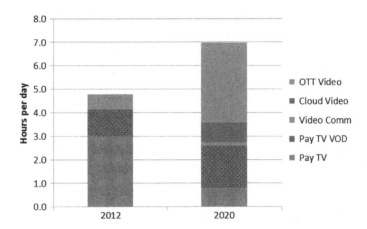

Figure 7.1 The ongoing transition from traditional TV programs (linear TV) to Internet-delivered video. Note: *OTT refers to over-the-top—in this case video, typically from an Internet content provider that rides over an operator's network—while VOD refers to video on demand.* Adopted from Market Realist.

This chapter details the challenges the Internet content providers face in expanding their data centers. This issue of data center scaling is not just the adding of more server racks and switches over time in a pay-as-you-grow fashion. For the Internet content providers, scaling implies the raw ability to reach the massive scale of computing they need. The basic networking architecture used to link servers and other equipment is described, as are its shortfalls.

Photonics plays an important role here in the form of pluggable optical transceivers, optical devices whose requirements continue to be pushed. But the demands of the data center are causing a more fundamental change in the use of photonics. The current partnership of chips and pluggable optics can only go so far, requiring photonics to be brought inside systems and closer to the chips. Moreover, there are already signs that bringing the optics onto the printed circuit board, closer to the chip, is itself insufficient longer term. The demands of the data center mean that optics, sooner or later, will be brought inside chips, or more accurately copackaged with them. Silicon photonics is the technology for such copackaging.

7.2 INTERNET CONTENT PROVIDERS ARE THE NEW DRIVERS OF PHOTONICS

Internet content providers may yet not spend as much as telcos annually on optical gear (as discussed in Chapter 5, Metro and Long-Haul Network Growth Demands Exponential Progress), but their fast growth and pressing requirements are beginning to shape the evolution of optical technology. The 100-Gb transceivers, now state of the art, are being purchased in considerable volumes for their data centers. The role of the optical transceiver in enabling the Internet content providers' businesses should not be underestimated.

The Internet content providers' optical communication requirements differ from those of the telcos. The Internet players are putting up huge data center complexes that can house more than 100,000 servers. The Internet content providers do not balk at using proprietary solutions if needed, and they will buy equipment from a sole supplier—practices eschewed by the telcos, which as regulated companies must adhere to industry standards and rely on several suppliers.

The Internet content providers are also more relaxed than telcos when it comes to equipment specifications. This is not because they have lower standards than the telcos. Rather, their equipment must meet less stringent demands given the controlled environment of the data center. The data center equipment also has a shorter life, with data center operators upgrading their systems as often as every 2 to 3 years. In contrast, telco equipment is deployed in harsher environments and must operate across a broader temperature range. Telco equipment can reside in the network for two decades or more.

As large-scale data centers can require hundreds of thousands of high-bandwidth optical transceivers, Internet content providers want inexpensive, power-efficient designs to minimize their costs. Such requirements represent a great commercial opportunity for silicon photonics, and not just because of bandwidth. Optics accounts for an increasing proportion of overall networking cost, so new technologies are being sought to better meet market needs.

The market for 100-Gb/s data center transceivers kicked off in the second half of 2016. To achieve this throughput, data is sent over several lanes requiring multiple lasers and receivers. Single-mode fiber is considered the best approach for future traffic growth due to its ability to support longer distances and higher bandwidth using wavelength-division multiplexing. But single-mode fiber and associated transceivers are relatively expensive when compared with links using multimode fiber.

As multiple active photonic components are required for each link over single-mode fiber, photonic integration—whereby multiple components in the transceiver are copackaged or even monolithically designed—promises to reduce costs. Silicon photonics is already being used to deliver such transceivers [3] (see Fig. 7.2).

It should be noted that the Internet content providers first requested 100-Gb transceivers in 2010. It took 6 years for the optical module makers to deliver products that addressed their needs. The industry's belated response reflects the traditional optical component market's development cycle. Now, with the growing role of the Internet content providers, things are changing. They demand much shorter development times and require products with a predictable delivery date. This is a role that silicon photonics, with its commonalities with the chip industry, can exploit.

Optical transceivers are but one part of what data center operators need. Silicon photonics' attractiveness stems from the fact that it can play a role across many levels of interconnect, a focus of this chapter.

Figure 7.2 Wafer of Luxtera's 100G PSM4 silicon photonics-based optical engines. From Luxtera.

7.3 DATA CENTER NETWORKING ARCHITECTURES AND THEIR LIMITATIONS

The challenge facing data center networking engineers is how to link efficiently an ever-growing number of servers. This is especially tricky given how tasks and data are distributed across the vast computing resources within a data center. Whereas, in a traditional campus-based local area network, a user request would typically be fulfilled by one server, today's data centers spread data sets and processing across multiple servers, requiring constant communication between them.

Such server traffic flows are referred to as East–West traffic, which means that the traffic stays within the data center, as opposed to North–South, which refers to traffic entering and exiting the data center. East–West traffic accounts for the bulk of the traffic flow within a data center. When a user logs into a social media site such as Facebook, e.g., the incoming request generates a near 1000-fold increase in traffic within the data center [4]. All the feeds unique to the user are aggregated, including what it calls anniversary posts—images or videos a user may have uploaded years ago—that have been archived and need to be retrieved from longer term storage.

To get a sense of the scale of Facebook's operations, hundreds of petabytes are stored in its data centers—a petabyte being one million gigabytes—while 1.79 billion users visit the site every month [5]. The world's population in 2016 is estimated to be 7.40 billion.

Scaling issues are not unique to Facebook. All the Internet content providers are grappling with how to ensure that they can keep growing servers, storage, and switching to meet their massive and growing individual operational requirements.

7.3.1 The Leaf-and-Spine Switching Architecture

Data center managers have adopted a standard hierarchical arrangement of network switches that they replicate across the data center to connect more and more servers and accommodate the traffic flows between them. A commonly used building-block switch architecture is known as leaf and spine [6].

Leaf and spine is a layered architecture. Leaf switches typically reside on the top of a server rack, hence they are also known as top-of-rack switches. The spine switches, typically having a higher switching capacity, link several leaf switches to enable them to talk to each other.

A data center manager wants the switch architecture to connect the end devices (e.g., typically servers, but it could also be storage units) as efficiently as possible. Efficiency refers to performance metrics such as latency—the time it takes for a data packet to travel from one server port to another—which ideally should be as low as practicable. Efficiency also means using the least amount of switch hardware to meet the networking goals: more switches mean more floor space, more power consumption, and more transceivers and cabling, all of which add cost. If a data center manager is going to use multiple leaf-and-spine arrangements to connect hundreds of thousands of servers, the basic leaf-spine unit had better be as efficient as possible.

Fig. 7.3 shows a simplistic leaf-and-spine arrangement to highlight how switch count and interconnect grow as more servers are added.

The basic leaf switch shown is a four-port device that can support up to four servers. Doubling the server count to eight increases the number of switches from one to six and sends the number of interconnects linking the switches up from nil to eight.

Fig. 7.4 shows the exponential impact of increasing the number of servers further. Note the interconnects grow more rapidly than the switch count. The cost of scaling a leaf-and-spine architecture is thus a

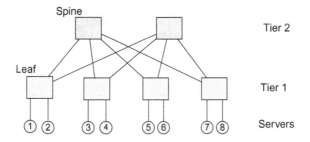

Figure 7.3 A simple leaf-and-spine architecture linking eight servers.

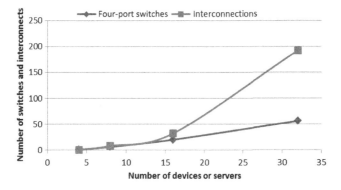

Figure 7.4 Impact of server count on the number of four-port switches and interconnects in a leaf-and-spine architecture.

function of the number of switches and the interconnects they use. From Fig. 7.4, what can also be learned is that an individual switch with a high port count is desirable.

Table 7.1 further shows the effect on a network of scaling the number of servers. The number of switch tiers needed increases, which complicates the cabling and increases cost. The number of servers that can be linked is determined by the switch port count in the spine. Adding servers to meet end-user requests require a significant increase in switches and ports—all expensive items.

In practice, data center networks are far more complicated than the example above. For a start, other networking elements are required besides servers and switches. One is edge routers that support Internet Protocol (IP) routing to handle external connectivity to the backbone and wide area network (WAN)—i.e., communication beyond just the

Table 7.1 The Number of Switches, Interconnects, and Switch Tiers Needed to Support End Devices in a Leaf-and-Spine Architecture			
Device Ports	**Four-Port Switches**	**Interconnects**	**Number of Tiers**
4	1	0	1
8	6	8	2
16	20	32	3
32	56	192	4

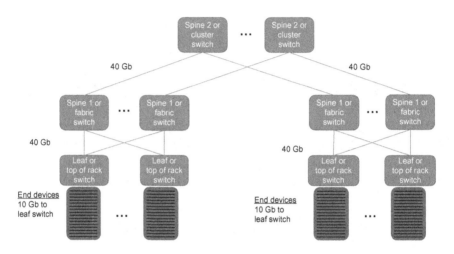

Figure 7.5 A three-tier leaf-and-spine architecture as described by Facebook. Modified from Facebook.

leaf and spine. Another is the load balancer that, as suggested by the name, has a view of workloads and ensures that one cluster of servers is not overworked while another is relatively idle.

Fig. 7.5 shows a schematic of a network described by Facebook [4]. Consider a 128-port leaf switch where 96 devices such as servers are connected using 10-Gb links. The remaining 32 leaf switch ports are used for uplinks. For example, ports can be grouped in units of four to create eight ports, each at 40 Gb/s, to connect the leaf switches to the spine, or fabric, switches. A three-tiered architecture is used in this example, with the second spine used to enable more end devices to be connected on the network.

Top-of-rack and spine switches, with 96 ports each, enable a total of 221,184 ports to be used to connect servers and other appliances. How is this number arrived at? The number of devices that can be attached is a function of the edge or top-of-rack port count, P_e, and the core or spine switches ports, P_c.

For a three-level design, the maximum number of devices or servers that can be connected is $P_e \times (P_c/2)^{(L-1)}$, where L is the number of layers [7].

The issue with such a topology, says Facebook, is the huge number of cables and optical transceivers needed. In this example, there are 442,368 links between the switches (where the number of links equals $P_c \times P_c/2$, as described above, where P_c is the number of core switch ports, in this case 96).

Not surprisingly, interconnect is a considerable part of the overall cost of scaling leaf and spine architectures. Moreover, the need for 100-Gb links and the need for single-mode transceivers affect the interconnect cost independently. It is this rising cost that is driving the need for cheaper interconnect.

The attraction of using the leaf-and-spine architecture is that it is scalable, predictable, and resilient. The switches are built using merchant silicon, and while these are complex chips, they are relatively inexpensive. Adding leaf switches and appliances does not require software changes as long as there are free spine ports. Multiple paths and spine-switch routing algorithms provide predictable transit times between the servers, ensuring good service for end users. And spine-switch failures can be circumvented with minor network performance degradation. All these factors are positives as far as data center managers are concerned.

The distance between the server and the top-of-rack switch is typically under 10 m, enabling copper cabling to be used. The link between the top-of-rack switch and the spine can be a few 100 m in length, but link lengths less than 50 m are typical. Although optics is used, the link cost is modest due to inexpensive optical transceivers and the short lengths.

As the network size increases and the number of servers and switches grows, so does the distance between the fabric (the spine 1 switches in Fig. 7.5) and the cluster (spine 2) switches, requiring more single-mode fiber and more costly single-mode transceivers.

Indeed, data center buildings can be distributed tens of kilometers apart to serve a metro area. Considering that redundancy is often desirable, such data center clusters can require a large number of links.

Microsoft talks of distances up to 70 km; optical engineers are developing solutions for this application. In this case, tiered spine switches need to talk to each other over such distances to make the servers in the distributed data center buildings appear as a single shared resource (see Section 5.5.1)

7.3.2 Higher-Order Radix Switches
One way to scale data center networking is to increase the number of ports—or radix—of the building-block switch. Using a high-radix switch reduces the overall interconnects, switch tiers, and switch boxes needed. This reduces overall equipment costs and the cost of operating them, as less power is consumed and less floor space is needed overall. A high-radix switch also benefits network performance by decreasing latency as fewer hops between switch stages are needed when two servers chat.

A market leader in 100-Gb switch silicon is Broadcom. The fabless chip company's flagship Ethernet switch integrated circuit (IC) is the StrataXGS Tomahawk with a switch capacity of 3.2 Tb: 32 ports at 100 Gb or 128 ports at 25 Gb [8]. While the bandwidth is higher than the top-of-rack switch shown in Fig. 7.5, the number of servers and other end devices supported is the same.

Ethernet switch equipment vendors have copackaged multiples of these switch chips to increase overall port count. Internally, these switches have a multitier CLOS, named after its inventor Charles Clos, a leaf-and-spine architecture, and the overall platform is typically used as a spine switch. But they are costly and power-hungry. Fig. 7.6 shows various vendors' 100-Gb high-port-count switches.

7.4 EMBEDDING OPTICS TO BENEFIT SYSTEMS

Although high-radix switches is one direction to support data center scaling as detailed above, current system data center requirements are resulting in a rethink of system design and architecture.

Research is being conducted as to how electronics and optics can be used to ensure more efficient data center networking performance. Such research is seeking to improve the data center computation and processing speed while reducing latency, cost, and power consumption. The resulting systems being proposed require new combinations of

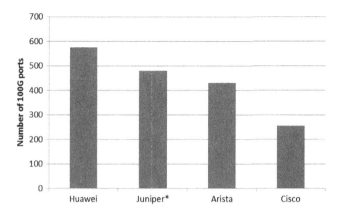

*Figure 7.6 A selection of the highest-port-count 100-Gb Ethernet switches available commercially. *Juniper announced, not shipping as of 3Q2016. Data from company reports.*

photonics with electronics, bringing optics inside systems and placed ever closer to the application-specific integrated circuit (ASIC). An example of such research is now discussed.

7.4.1 An Electronic-Optic Switch Architecture for Large Data Centers

Keren Bergman's career has always been at the intersection of computing systems and photonics, first with high-performance computing and now data centers. A professor of electrical engineering at Columbia University, Bergman's current focus is on how photonics can improve overall computational performance of the data center while minimizing energy consumption.

A key challenge Professor Bergman and her research team are tackling is how photonics can best link 100,000 or more servers. Instead of developing a large high-radix—320 or even 1000-port—optical switch that would connect to the top-of-rack or leaf switches to offer an optical complement to the existing electrical switching architecture [9], Bergman and her team are taking a different approach: embedding photonics inside the electronic switches. Such an approach will be enabled by silicon photonics, says Bergman: "That is key."

Bringing silicon photonics onto the switch chip board will create a new entity, a combination of electrical routing and optical switching. This results in the best of both worlds: all the benefits of an electronics switch chip such as programmability and packet buffering, with much denser optical input–output enabled by silicon photonics. In turn, by adding silicon photonics switching, traffic will be sent either through the electrical or the optical switching.

Embedded photonic switching offers a fundamentally different way of moving data across the data center, says Bergman. Photonics lends itself to data being broadcast to multiple locations—a networking approach commonly used for data packets known as multicasting—while individual transmitted wavelengths carrying data can be picked off at locations as required. Multicasting is much harder to do electrically and is costly in terms of latency and energy, especially over data center spans.

These optical-enabled data-transfer techniques offer a powerful way to deal with the huge growth in data traffic transiting the data center. Applications such as data analytics, used to extract valuable information from large data sets, is generating vast amounts of traffic between servers. Bergman sees a direct link between data analytics, which has great commercial value to companies, and embedded optics: "This is very recent, only in the last few months have we converged on this in our research."

One goal of the team's work is that a data center manager will not even know that the networking systems include embedded optics; the system will look and be controlled as current systems are today except it will run faster and consume less energy. This is extremely attractive to data center managers, whereas the adding of a large external optical switch requires them to change the data center controller software, she says.

But once silicon photonics becomes embedded within such systems, things will change. The resulting bandwidth of the networking architecture will become much higher and more uniform. Then, the need for hierarchical switching architectures will start to diminish. Professor Bergman predicts that we will see differently architected data centers in the next 3 to 5 years.

Silicon photonics start-up Rockley Photonics is also developing a switch architecture for the data center based on optics and a custom ASIC. As of this writing, Rockley is still in stealth mode. But it has said that its switch concept will scale with Moore's law, meaning it will support a doubling of capacity every 2 years [10].

The idea of embedding optics in systems is not new. Intel announced a disaggregated server architecture enabled using silicon photonics interconnects incorporated within the system. And start-up Compass Electro Optical Systems (Compass EOS) went a step further and copackaged optics with electronics to create a complex ASIC with optical input–output. While both designs faltered commercially, they provide valuable case studies in highlighting the potential of bringing optics inside systems. Both systems are now described.

7.4.2 Intel's Disaggregated Server Rack Scale Architecture

Intel introduced its Rack Scale Architecture in 2013, a disaggregated server design enabled by silicon photonics. A disaggregated design physically separates the various parts of a server—processors, memory, storage, and switching—linking them using high-speed, low-cost interconnects instead [11].

Using such an approach, the server's performance can be tailored to a given task. If more memory is needed, it can be added without upgrading the entire server. And as the elements making up the server have different upgrade cycles, when a more powerful processor is released, it can be slotted in without having to upgrade all the other elements. This can extend the life of the platform to as long as 20 years [12]. There are also environmental cost benefits. If the processors generate more heat than flash memory, they can be dealt with as a microenvironment and cooled appropriately (see Fig. 7.7).

However, the disaggregated server design only works if the various elements can be connected with sufficiently low latency, and that requires high-speed, cheap interconnect. Intel chose to use silicon photonics. The silicon photonics transceiver operated at 1310 nm and was coupled to multimode fiber. Intel worked with a leading fiber manufacturer Corning to develop the multimode fiber and demonstrated it transmitting at 28 Gb/s over nearly a 1-km distance [13], although distances of a few 100 m are sufficient for these applications.

Intel also worked with US Conec to develop a custom connector that had high bandwidth and was robust. The resulting connector used a fiber ribbon, 16 strands wide and 4 rows high, yielding a cable, called the MXC, with 64 lanes, each at 28 Gb/s, for a total duplex bandwidth of 1.6 Tb/s [14].

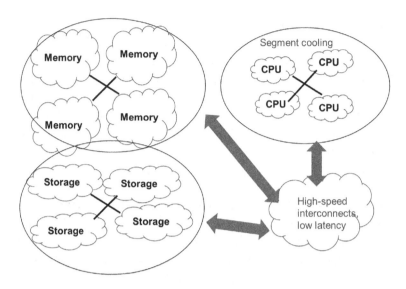

Figure 7.7 Device-specific environmental conditioning enabled by disaggregation.

Intel developed an expanded beam connector to ensure the MXC's ease of use. This enabled robust coupling between the light source and the fiber strands ensuring a tolerance to dust and dirt, the cause of most optical link failures.

By choosing multimode fiber and developing a low-loss, terabit-plus connector, Intel hoped to offer a novel server architecture for the data center with silicon photonics at its core. However, in early 2015, Intel announced a delay in its silicon photonics modules and said they would ship by the year's end [15]. In August 2016, Intel did indeed announce silicon photonics modules, but not for the Rack Scale Server. Rather, these are pluggable 100-Gb PSM4 and CWDM4 QSFP28 transceivers for switch equipment in the data center [16].

So where does that leave Intel's Rack Scale Architecture?

According to an Intel spokesperson, silicon photonics is not being used in Intel's Rack Scale design, primarily because optical connectivity is not required at the data rates and distances used inside a rack today. Copper continues to push back the technical boundaries and keeps an all-optical interconnection the subject of "future" plans. But the spokesperson added that as server resources disaggregate, and throughput and distance requirements increase beyond the capability of current interconnect technologies such as copper and vertical-cavity surface-emitting lasers (VCSELs), then silicon photonics is expected to provide the high-speed optical connectivity between a new generation of pooled resources.

So despite Intel's technological heft and the design's development effort and ambition, the company concluded that a silicon photonics design is premature for this application. Moreover, copper already enables such disaggregated designs. Equipment makers such as Cisco Systems and Dell have server disaggregation products that use copper-based high-speed connections. Ericsson has also announced a disaggregated hardware system based on Intel's Rack Scale Architecture where the connectivity is optical but it is not based on silicon photonics [17].

7.4.3 Compass-EOS: Copackaging Optics and Silicon

Compass-EOS was arguably the first company to offer optics copackaged with a complex chip. The company, which later became Compass Networks, provides a valuable case study from a technology and commercial perspective.

The ambitious Israeli start-up developed an IP core router to compete with the likes of Cisco Systems, Juniper Networks, Alcatel-Lucent (now Nokia), and Chinese giant Huawei.

It developed a way to integrate optics with its complex traffic manager chip design, resulting in a simpler and lower-power optically enabled core router platform. However, despite the novel chip, the company ultimately failed commercially, largely because its software team of 60 engineers could not compete with its much larger IP core router rivals.

Simply put, a router is a networking platform at the heart of the Internet. It takes IP traffic in the form of packets on its input ports and forwards them to their destination via its output ports. To do this, two chip types are used: a network processor and a traffic manager.

The network processor does all the packet processing—it takes the packet's header and uses a lookup routing table to determine its destination and inserts a new updated header frame in the packet.

The traffic manager oversees billions of packets. The chip implements the queueing protocols and, based on a set of rules, determines which packets have priority on what ports. In a conventional IP router, there are also switch fabric chips that send the packets between the cards to the right router output ports.

Compass-EOS designed its router between 2007 and 2008 and used a merchant chip for the network processor. But it designed its own complex traffic manager ASIC and added a twist by figuring out a way to add optics to the chip. As a result, no switch fabric was needed. Instead of the traffic manager going via the router's electrical backplane to a traffic manager on another card, each optically enabled traffic manager had sufficient bandwidth to connect to all the other traffic managers. Eight traffic manager chips in total were used on four line cards—all linked in a fully connected optical mesh.

As lead engineer Kobi Hasharoni put it, there was no backplane, which is why the Compass-EOS routers were so much more compact than those of its competitors.

At the time of the design, Compass-EOS did not consider using silicon photonics, which was deemed too immature. Instead, Compass-EOS used its ingenuity to figure out how to couple multiple VCSELs and photodetectors onto the chip.

VCSELs in 2007 were at 10 Gb/s, and Compass-EOS chose to operate them at a more relaxed rate of 8 Gb/s. In total, 168 VCSELs and 168 photodetectors were used on the traffic manager chip, enabling 1.344 Tb of traffic for transmit and the same for receive. The resulting chip with optical input–output consumed one-fifth of the power of a chip using electrical-only connections and achieved a 12-fold bandwidth-density improvement compared with electrical for the same chip area. Compass-EOS was developing a second-generation chip with a 16-Tb input–output before the project was canceled.

Hasharoni says that were he and his hardware colleagues to tackle a similar design today, they would use silicon photonics instead of VCSELs. The design would better support single-mode fiber, and the packaging would be easier with both the ASIC and optics being silicon.

What Compass-EOS demonstrated in 2010—arguably at least a decade ahead of its time—is how optics can be integrated alongside a complex chip to benefit the system architecture. In this case, it resulted in a more compact, lower-power IP router that was less costly to operate [18]. But despite all the router's hardware ingenuity, the venture ultimately failed.

The start-up's experiences were different from Intel's with its Rack Scale Architecture. Compass-EOS' hardware was a more ambitious copackaged chip, and the company did sell its IP router to several leading telcos. But the technical benefits of integrated optics within a system did not guarantee commercial success.

Before discussing the particular component technologies that will enable wide adoption of optics, first on the printed circuit board, then copackaged with silicon, the issues associated with pluggable optical modules are the subject of the next section.

7.5 DATA CENTER INPUT–OUTPUT CHALLENGES

Fig. 7.8 shows the evolution of optics: how it will move from the faceplate of systems in the form of pluggable modules to the printed circuit board

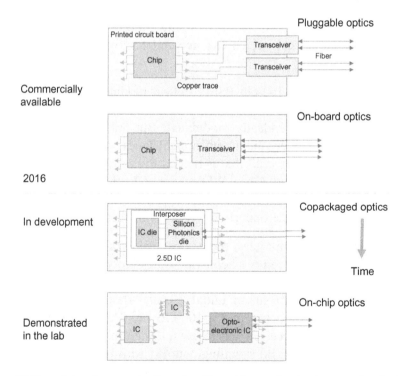

Figure 7.8 The evolution of system optics: From pluggables modules to on-board optics to copackaged optics to on-chip optics.

and then will be combined with silicon in the same package. Ultimately, optics and complementary metal-oxide semiconductor (CMOS) circuitry will coexist on the same die. The wider issues associated with these four approaches are discussed in the remaining part of the chapter. We start with the current approach based on pluggable optical modules.

7.5.1 Pluggable Optical Transceivers Evolution: Photonic Integration for 100-Gb

The 100-Gb transceivers being adopted to connect leaf and spine switches, and between the spine switches, use multiple data lanes, each lane having a transmitter and receiver. The main two approaches are to use wavelength-division multiplexing or parallel optics, as described in Appendix 1, Optical Communications Primer. For the former, four lasers are used to generate 4 wavelengths, each at 25 Gb/s, which are multiplexed onto a single-mode fiber, while with parallel optics one laser can be shared to generate the four physical transmit and four receive lanes, each at 25 Gb/s.

The cost of a laser is important. A single-mode transceiver is dominated by the laser cost, which can account for over 30% of the total bill of materials.

The parallel optics 100-Gb PSM4 optical module specification has a cost advantage because it can share a single laser across all four lanes, each lane being independently modulated. But link cost increases with distance because a fiber ribbon cable, eight lanes wide, is used. In contrast, a single fiber pair is all that is needed for wavelength-division multiplexing using the 100-Gb CWDM4 or CLR4 modules. Photonic integration is the only technological approach that provides optical module vendors with the opportunity to meet the demanding price goals.

For these 100-Gb modules, there is ostensibly no difference in optical performance between a silicon photonics-based transceiver design and an indium phosphide one. But like what happened with 10-Gb devices, which went from an initial parallel design to a serial one, developers are plotting to build a 100-Gb serial design that uses one transmitter and one receiver only. Such a design requires increasing the electrical lanes to 50-GBd signaling coupled with the multilevel signaling scheme of four-level pulse-amplitude modulation (PAM4) that allows 2 bits to be transmitted per symbol [19]. PAM4 is becoming an important modulation scheme to address evolving data center requirements.

Moreover, 100-Gb is the latest but not the last speed stop in the data center. Internet content providers are already eyeing 200- and 400-Gb links and even 1-Tb ones. All these designs require photonic integration and high-speed electronics that will benefit from a close coupling between the optics and electronics, whether the optics are embedded on the printed circuit board near the chip or are copackaged with the chip.

Using an advanced CMOS packaging process for silicon photonics provides the most promising path for high performance and low cost.

7.5.2 The Move to Single-mode Fiber

Both single-mode and multimode optical transceivers are used in data centers. Traditionally, links from top-of-rack switches to spine switches use multimode fiber while links between spine switches use single-mode fiber. Microsoft is one Internet content provider that plans to use only single-mode fiber for all its future data centers. Many other Internet content providers are following Microsoft, but enterprise players, e.g., continue to use multimode.

Multimode fiber links are the cheapest because they use inexpensive lasers and hence inexpensive transceivers based on VCSELs. However, with each increase in data rate, the transmission distance standardized over a multimode fiber diminishes. This, in part, explains Microsoft's decision to adopt single-mode fiber, to support long spans within its data centers, to simplify its networking decisions and future-proof its fiber investment as data rates inevitably rise. Using only single-mode fiber also reduces the module types that a data center manager must keep as inventory and hence costs.

Data center managers' adoption of single-mode fiber is good news for photonic integration and for silicon photonics because it increases the overall volume for single-mode transceivers, essential for reducing costs.

7.5.3 Pluggable Transceivers for the Server—to—Top-of-Row Switch Optical Opportunity

Current links between servers and top-of-row switches are at 10 Gb/s because that is the link speed needed by the servers used by the largest

Internet content providers. Copper cabling is commonly used to connect the two given the link distance is typically is under 3 m.

Optical cables have started replacing copper for such links as costs have come down. And optics beats copper when it comes to power consumption. Comparing optical to electrical switches in, e.g., a 10-Gb port, the power consumption of copper is 4−6 W, whereas for optics it is 0.5 W [20].

Another trend benefiting the switchover to optics is that as transmission speeds increase, the distance supported by a copper link decreases. As network interface cards for servers move to 25 Gb/s, copper can support up to 3-m link length. But moving to 50-Gb and then 100-Gb links, copper's reach may become too short for server-to-leaf switch connections.

Optics will play an increasingly important role connecting servers to switches. Here, silicon photonics is battling multimode VCSEL-based links. For silicon photonics to replace VCSELs, it must match VCSELs' optical performance and low power consumption and be lower cost—a challenging set of requirements. Silicon photonics has shown it can be competitive with VCSELs when used in active optical cables.

7.5.4 Transition One: From Pluggables to On-Board Optics

Given how switch capacity will continue to grow, a key issue that needs to be addressed is increasing the capacity that can be supported on the faceplate of platforms. The platform itself will not get any larger so what is required is to pack more, higher-capacity pluggable modules on a switch's front panel.

The market currently uses 100-Gb pluggable modules, and the smallest optical transceiver package available for 100-Gb is the QSFP28 form factor. Thirty-two QSFP28s are typically used on the front panel matching the 3.2 Tb of capacity of current state-of-the-art switch chips, and this can be stretched to as many as 36 QSFP28 modules.

Mellanox is one equipment company that is confident it can get to 200 Gb/s in a QSFP28 module using its silicon photonics technology. The module is expected in 2017 and will enable a faceplate density of 7.2 Tb/s using 36 200-Gb modules [21].

According to Mellanox's Mehdi Asghari, the company's 25-Gb silicon photonics modulator and photodetector can already operate at 50 Gb/s. This means it can use the simpler nonreturn-to-zero modulation scheme rather than PAM4.

The challenge using nonreturn-to-zero signaling is getting the associated electronics to work at 50 Gb. This is a complex radio frequency design challenge that requires skilled analog design engineers. The design challenges include getting the drive electronics, the assembly, the wirebonds used to connect the chip to the packaging, and the printed circuit board tracks to all work with good signal integrity at 50 Gb. The benefit of going from 25 to 50 Gb/s using nonreturn-to-zero is that there is less signal loss than using PAM4 [22]. Using PAM4 also introduces extra latency. Nonreturn-to-zero is always the best option if you can do it, says Asghari.

The next-generation data rate is 400 Gb/s, and the proposed pluggable optical transceiver packages to house the optics are the quad small form-factor pluggable double density (QSFP-DD) and the μQSFP. Fig. 7.9 highlights these new pluggable form factor options being developed for next-generation switches as summarized by the industry organization, the Ethernet Alliance [23].

The QSFP-DD uses eight lanes that can operate at 25 Gb using nonreturn-to-zero modulation and 50 Gb using PAM4 to

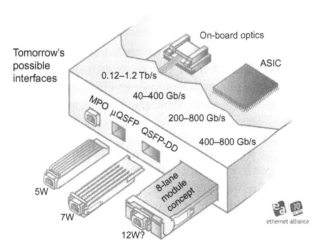

Figure 7.9 Solutions being developed for next-generation switches. From Ethernet Alliance.

support 200- or 400-Gb ports. Thirty-six modules can be packed on a one-rack-unit front panel, for an aggregate bandwidth of 14.4 Tb.

The μQSFP is smaller—approximately, the same size as the small form-factor pluggable (SFP) pluggable module. As in the QSFP, there are four electrical channels, each lane using 25-Gb nonreturn-to-zero signaling. The multisource agreement is intended to support 50-Gb using PAM4 as well. While operating at 25 Gb, an aggregate one-rack-unit faceplate density of 7.2 Tb is achieved using 72 ports.

Optical transceiver vendors are well on the way to defining and preparing the next high-bandwidth transceivers with a small enough size to support next-generation 12.8-Tb switches. Internet content providers are already clamoring for these switches and associated optical interconnects.

One way to tackle the issue of limited faceplate density is to adopt on-board optics. Embedded, mid-board or on-board optics moves the optical transceiver off the front plate and onto the printed circuit board, closer to the ASIC.

Several advantages result. First, the high-speed electrical signal path from the ASIC to the optics is shorter, simplifying the printed circuit board design. And using on-board optics only, a fiber connector is fed to the front of the equipment. This means the number of transceivers depends on how many embedded optics modules can be fitted on the board, not how many pluggable modules can be fitted on the equipment's front panel. Freeing the front panel of pluggable modules also allows for more ventilation holes, improving the air flow used to cool the equipment.

However, on-board optics has its own issues. One is that all the switch slots are populated from the start with optics, something that can be avoided with pluggable modules. This may be acceptable for the Internet content providers, but many enterprises prefer a pay-as-you-grow model.

Another is that if an optical link fails, the entire board has to be replaced—an expensive proposition. Solutions lie in optical redundancy, in understanding the failure mechanism, and in design and planning, such as making the optical modules field-replaceable.

There is also no industry standard for embedded optics. COBO, the Consortium for On-Board Optics, is an industry initiative backed by

Microsoft that is working to develop the first standard solutions for embedded optics [24]. COBO is working on 400-Gb/s designs and a way of combining two side-by-side to deliver 800-Gb/s interfaces. The embedded optics will support multimode and single-mode fiber over different reaches. COBO is considering enabling mid-board optics to support distances of 100 km using coherent technology, as discussed in Section 5.5.1.

Luxtera is providing 2-by-100-Gb PSM4 embedded optics modules in volume to several equipment makers. Luxtera is also a member of COBO, as is Mellanox. Yet while both companies welcome the advent of a standard, they admit there is an industry caution among companies about embracing on-board optical technology.

Luxtera cites the issue of field serviceability as one concern customers have. "Front faceplate pluggable modules are much easier to replace but come with a higher implementation cost," says Luxtera's Peter De Dobbelaere. He claims that the greater reliability of silicon photonics when used for embedded optics also helps allay customers concerns.

Mellanox says that the key issue is the input−output bottleneck on the ASIC and embedded optics, while increasing the input−output of equipment, does nothing to address the chip's pinch point, as is explained in the next section. Asghari also warns not to dismiss pluggable optics which he expects to continue to evolve for at least two more generations.

The use of silicon photonics suits mid-board optics design due to the small size, low power, and low cost needed. The potential for high volumes is also attractive. However, embedded modules can use any optical technology as long as it provides adequate performance. And while silicon photonics may be well suited, it is up against traditional VCSEL technology for short-link lengths and indium phosphide−based offerings for longer reach ones.

The advent of on-board optics is an important milestone in the development of photonics in the data center. Brad Booth, the chair of COBO, says that the deployment of technology in systems will help the industry learn what embedded optics brings and what some of the challenges are. There is no revolutionary change that happens with technology, he says, it all has to be evolutionary. In other words, embedded optics represents an important stepping-stone for the next development—and the most significant development for silicon photonics—that of copackaged optics.

7.5.5 Transition Two: From On-Board Optics to Copackaged Optics

The introduction of higher-capacity switch chips is driving pluggable module developments and a need for optics inside systems.

Current generation switch chips support 3.2 Tb of capacity, and the 6.4 Tb Tomahawk II from Broadcom, at the time of writing, is already sampling. The QSFP28 pluggable module developments will be able to support these next-generation chips, as discussed in the previous section. The next switch chip advance—to 12.8-Tb, a capacity already sought by the Internet content providers—is expected in 2018. Here too, next-generation pluggables such as the QSFP56, QSFP-DD, and the μQSFP will deliver sufficient input–output capacity at the system's front panel to support such switch silicon.

But faceplate density—having sufficient input–output capacity to support the faster switch silicon—is only one of the system design issues. To go from 3.2 to 6.4 Tb, the input–output of the switch chip must double. Broadcom's 6.4 Tb switch chip crams 256 25 Gb/s signals on-chip by using a more advanced CMOS process node.

Each of these 25-Gb input–output signals is generated using a serdes (serializer/deserializer) circuit. The role of the serdes is to serialize the switch data and drive it across electrical lines on the printed circuit board to the pluggable optical transceiver. Clearly, going from 3.2 to a 6.4-Tb chip doubles the number of 25-Gb lines driving signals across the printed circuit board to the 100-Gb pluggable modules. Inside the optical module, there is a retimer circuit that cleans up the received electrical signal before being passed to the optics for transmission.

According to Luxtera's De Dobbelaere, switch vendors have to sometimes put an additional retimer chip between the switch ASIC and module to address signal integrity issues associated with longer printed circuit board traces. Retimer circuitry adds to the board's design complexity and dissipates power, and the problem only worsens as channel count and data rate go up.

And of course that is just what will happen with the advent of 12.8-Tb switch silicon. To achieve 12.8 Tb, two approaches are possible: either go to an even more advanced—and expensive—CMOS process node such as 7 nm to get the serdes to 50 Gb/s or continue to use 25-Gbd/s serdes but add PAM4 to double the switch capacity. According to De Dobbelaere, the industry has settled on 25 GBd and PAM4 for now. Note, whichever approach had been chosen, it would only exacerbate the chip's power consumption—the serdes are expected to consume over half the chip's total power—and the issue of retimers.

This is where using on-board optics close to the switch chip helps. The shorter traces between the chip and the on-board optics consume less power. But embedded optics does nothing to solve the fundamental problem associated with switch ASICs: the finite number of input–output lines a chip can support.

Simply put, the chip's high-speed input–output signaling must be placed on the chip's edge due to printed circuit board design considerations. The chip's perimeter is finite, and the input–output signals must be spaced a certain distance apart. The serdes circuitry connects to the printed circuit board through little balls on the chip packaging's base, an arrangement known as a ball grid array. There are only so many balls that can be supported on a chip, and only so much of the ASIC's area that can be used for input–output without exceeding the maximum chip size.

The next switch chip after 12.8 Tb will be 25.6 Tb, scheduled for 2020. For the chip to support such a capacity, 100-Gb signals will be required using 50-GBd and PAM4. Such a solution will burn an even greater proportion of the chip's power, complicate the packaging and printed circuit board designs, and diminish the transmission distances possible over the board's traces. On-board optics simplifies the signaling between an ASIC and the on-board optics and lowers the power consumed, but it cannot address the chip's input–output bottleneck.

What is required is a way to expand the number of input–output signals off the chip and simplify the drive requirements on the individual serdes. This is where interposer technology—developed by the chip industry—can help. By using interposer technology to copackage the ASIC with silicon photonics, both these goals are achieved.

The semiconductor industry is adopting interposer technology as part of its work developing 2.5D- and 3D-packaging technology. The chip industry has long used 2D-packaging techniques whereby more than one chip die—the bare silicon chip before it is packaged—is bonded onto a common substrate. The substrate material includes laminates (a form of printed circuit board with fine copper lines), ceramic, or silicon. 2D packaging is known by a variety of names such as system in package and multichip module. The advantage of 2D packaging is the ability to combine different chip technologies in one package. This benefits yield and allows device evolution without having to redesign a single more complex chip [25].

The 2.5D chip extends the concept by adding a silicon interposer between the different dice and the substrate, as shown in Fig. 7.10. The interposer is a slice of silicon which has metal traces on both its surfaces. The chips are bonded to the upper surface while the lower surface is connected to the common substrate using standard flip-chip bumps. Through silicon vias (TSV) technology is used to connect the two metal layers to allow the dice to communicate with the printed circuit board. For completeness, 3D packaging extends the concept by having dice stacked on top of each other and which are connected to the interposer.

What Fig. 7.10 shows is that the chip dice are connected to the interposer's upper surface using micro-bumps. These are a tenth the size of the flip-chip bumps and the bumps used by the ball grid array. And this is the key benefit: the interconnect on the interposer allows for much greater interconnect bandwidth between the dice, enabling a tremendous increase in capacity and device performance.

For silicon photonics, the technology holds huge potential. By adding a silicon photonics die on the interposer, an additional input—output route using fiber off the 2.5D chip becomes available.

Using 2.5D technology, silicon photonics vendors can thus benefit from yet another technology that has been developed by the chip

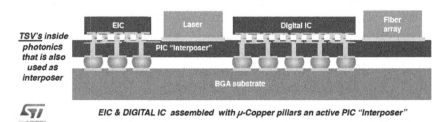

Figure 7.10 Photonic integrated circuit interposer. TSV is a through silicon via, a vertical electrical connection approach, EIC refers to an electronic integrated circuit, PIC is photonic integrated circuit, and BGA is a ball grid array. From © STMicroelectronics. Used with permission.

industry. Using such a technology, designers can develop designs similar to the ASIC with optical input—output pioneered by Compass-EOS.

For the data center, 2.5D packaging offers a solution that overcomes the input—output bottleneck of the switch chips, something that optical pluggable modules and on-board optics cannot address. It is also a technology that silicon photonics is inherently able to exploit. This is not the case with the established optical technologies.

And it doesn't stop there: high-performance computing processors and future field-programable gate arrays (FPGAs) can benefit from being copackaged with silicon photonics. An FPGA is an important chip, with the more advanced versions containing huge numbers of logic gates as well as 25-Gb/s serdes that can be customized to implement datacom and telecom chip designs [26].

Professor Bergman points out that her team's research work includes ways to embed photonics within electronic switches to improve networking performance while reducing the energy required, as discussed in Section 7.4.1. This will be enabled using 2.5D integration technology, and silicon photonics allows the embedding of optical input—output right next to the electronics switch chip.

Another part of her data center work is looking at computing nodes. Such nodes comprise general multicore devices, specialist graphic processing units, and high bandwidth memory. Chip-to-memory communications are becoming a key pinch point in these systems and are ripe for the use of interposer technology.

Bergman points out that only so many dice can fit on an interposer but that the bandwidth that silicon photonics brings means that 2.5D packages can be linked. "The really beautiful thing with photonics is that you add more interposers because the bandwidth off that 2.5D chip is just as high as on the chip," she says.

Silicon photonics luminary Professor Lionel Kimerling, at MIT's Department of Materials Science and Engineering, agrees that board-level integration is an important area of focus and the question is what solution will scale for the next 20 years. The chip industry will not make a big investment in something that will not scale for more than a couple of generations, he says.

In his view, the chip is becoming almost irrelevant given it will no longer scale with the demise of Moore's law. "So it is all about how can I scale the number of chips in a package," he says. This is promoting system-in-package technology. Once this starts, there is no doubt that there will need to be optical interconnect coming out of such a package, he says: "And no one really doubts that before long, you will need optical interconnect between the chips inside the package."

This is the potential interposer technology offers. But combining photonics and electronics chips using an interposer is not without challenges. The semiconductor industry itself has only recently started to embrace the technology and adding photonics presents its own issues. Thermal management and where to place the laser being two major ones designers must resolve.

Will companies pass on on-board optics and adopt a 2.5D design when they transition their systems away from pluggable optical transceivers? It is possible but a more likely path is that the industry will embrace on-board optics first. New technologies take time, and there is much to be learned from embedded optics before copackaged optics will be ready and the industry will be willing to make the jump.

Meanwhile, startups are now becoming active in this area.

Ayar Labs is a US start-up that is developing a 3.2-Tb optical interconnect chip designed to sit alongside switch silicon. The chip aims to replace 32 100-Gb pluggable modules or, in future, eight more complex 400-Gb modules, and it is expected to emerge starting in 2018.

Another start-up developing a silicon photonics device to sit alongside chips in the data center is Sicoya from Germany. It is targeting servers first but says the technology can be used for other equipment in the data center such as switches and routers. It says its silicon photonics device is designed for chip-to-chip communications and could be placed very close to the processor or even copackaged in a system-in-package design [27].

7.6 ADDING PHOTONICS TO ULTRALARGE-SCALE CHIPS

The benefits of copackaging include higher-performance chips, lower-power consumption, and greater signal integrity although it will require considerable development work before it is commercially deployed in volume.

Will an optics and photonics union end at copackaging? Extrapolating, could not optics end up in the chip, not just bolted alongside? Integrating optics within the chip would enable optical communications inside the chip and to other such chips.

A group of US academics in a paper published in the scientific journal *Nature* have demonstrated just that: a microprocessor that integrates logic and silicon photonics in one chip, with the optics enabling communications between chips [28].

Vladimir Stojanovic, one of the academics involved in the project and based at the University of California, Berkeley, claims it is the first time a microprocessor has communicated with the external world using something other than electronics.

The chip features two processor cores, as shown in Fig. 7.11, and 1 MB of on-chip memory, and comprises 70 million transistors and 850 optical components. The chip is also notable in that the researchers achieved their goal of fabricating it on a standard IBM 45-nm CMOS line without any alteration. They managed to implement the photonics functions using a CMOS process tuned for digital logic— what they call "zero-change" silicon photonics.

Pursuing a zero-change process was initially met with skepticism, says Stojanovic. People thought that making no changes to the process would prove too restrictive and lead to very poor optical device performance. Indeed, the first designs didn't work. But the team slowly mastered the process, making simple optical devices before moving onto more complex designs.

The chip uses a micro-ring resonator for modulation, while the laser source is external to the chip. The modulator, which is known to be sensitive to small temperature changes, is corrected with on-chip electronics.

The team has demonstrated two of its microprocessor chips talking to each other. One processor talked to the memory of the second chip that was 4 m away. Two chips were used rather than one—going off-chip before returning—to prove that the communication was indeed

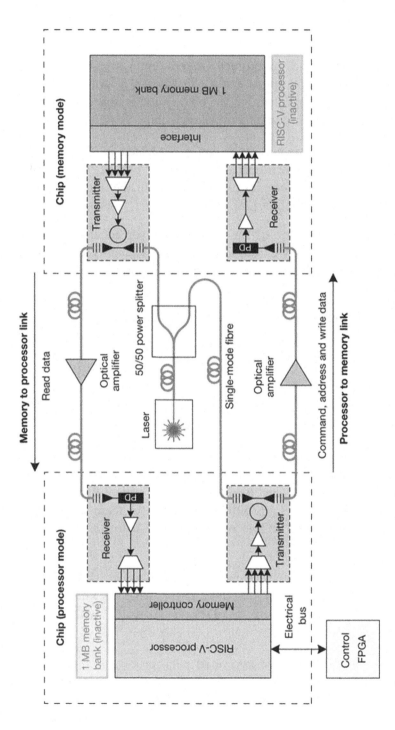

Figure 7.11 Optical interconnection integrated with microprocessors. Reprinted by permission from Macmillan Publishers Ltd: Chen Sun et al., Single-chip microprocessor that communicates directly using light, Nature 528, 534–538 (24 December 2015), copyright 2015.

optical as there is an internal electrical bus on the chip linking the processor and memory.

For the demonstration, a single laser operating at 1183 nm feeds the two paths linking the memory and processor. Each link operates at 2.5 Gb/s for a total bandwidth of 5 Gb/s.

However, for the demonstration, the microprocessor was clocked at one-eightieth of its 1.65-GHz clock speed—20.7 MHz—because only one wavelength was used to carry data and a higher clock speed would have flooded the link.

The microprocessor design can support 11 wavelengths for a total bandwidth of 55 Gb/s, while the silicon photonics technology itself will support between 16 and 32 wavelengths overall.

The group is lab-testing a new iteration of the chip that promises to run the processor at full clock speed. The factor-of-80 speedup is supported by using 10 wavelengths instead of one, each at 10 Gb/s, while the design will support duplex communications. The latest chip also features improved optical functions. "It has better devices all over the place: better modulators, photodetectors, and gratings. It keeps evolving," says Stojanovic.

The microprocessor development is hugely impressive. The demonstration is in effect two generations ahead of what's being developed today with transceivers. But the work already shows two things. One is that on-chip photonics can work alongside complex logic in the form of a two-core processor and memory. And the researchers achieved the photonics design without altering a standard CMOS process.

Optical communications using silicon photonics inside a chip is a long way off. The researchers deliberately chose to demonstrate chip-to-chip communications because they recognize that is the next big opportunity. As Stojanovic says, that is where the biggest bang for the buck is.

7.7 PULLING IT ALL TOGETHER

Data centers are factories of profit for the Internet giants. Such "computer buildings" require scalable networks to connect a huge and growing number of servers. The networks need to be predictable, power-efficient, and cost-effective. Silicon photonics is best suited to support these demands.

Reducing optical transceiver cost is the first identifiable opportunity for silicon photonics. Transceiver cost is a key issue given that the number of interconnects is growing exponentially as servers are added to the network.

Silicon photonics is a good candidate technology for optical transceivers for other reasons too. The Internet content providers need 100-Gb single-mode transceivers to support their hyperscale data centers, and such designs need photonic integration. That suits silicon photonics, which is also single-moded, thereby matching end-customer needs.

Delivering low-cost transceivers is one approach to enable highly scalable data centers. Another is to make high-radix switches. But what is really required is bringing optics inside the switch platform, to help switch chips grow their port count and the ports' speed, and to embed complementary optical switching to further enhance overall performance and reduce the energy used.

This is leading to optics moving off the faceplate and onto the board hosting the switch chip. And from there, bringing the optics even closer to the switch silicon using 2.5D copackaging technology. Silicon photonics is being used for pluggables and can play a role for embedded, but it comes into its own with copackaging.

Silicon photonics has even been shown working within a complex multicore microprocessor. The market may not be ready for all these solutions, but silicon photonics has already crossed the finishing line.

Key Takeaways

- The leaf-spine switching architecture has become the default approach to link servers, but as the number of servers grows, the number of switches and the interconnect between them grows exponentially.
- As data rates grow the trend is towards single-mode integrated transceivers, a shift that benefits silicon photonics. But to get better network scaling efficiencies, higher-capacity radix switches are needed.
- To improve systems performance in data centers, optics is being moved closer to the electronics. A disaggregated server using silicon photonics

and a novel IP router have already demonstrated such benefits. But these examples were not successful commercially, partly because they were too early for the marketplace.

- On-board optics and copackaged optics alongside silicon represent key milestones in the evolution of silicon photonics that promise to benefit many of the systems used in data centers: server nodes, switch silicon, networking platforms and storage.
- The end game is optics and electronics in one chip. This has already been demonstrated with optics piggybacking on a standard CMOS process. We can think of no better example that shows the long-term potential of silicon photonics.

REFERENCES

[1] Munroe R. Thing explainer: complicated stuff in simple words. London, John Murray; 2015. pp. 65.

[2] Why did Juniper acquire BTI systems, < http://marketrealist.com/2016/04/juniper-acquire-bti-systems/ > ; April 2016.

[3] Intel's 100-gigabit silicon photonics move. Gazettabyte, < http://www.gazettabyte.com/home/2016/8/21/intels-100-gigabit-silicon-photonics-move.html > ; August 21, 2016.

[4] Farrington N, Andreyev A. Facebook's data center network architecture. In: Data center network engineer, IEEE optical interconnects conference (Santa Fe, New Mexico), < http://nathanfarrington.com/papers/facebook-oic13.pdf > .

[5] Zephoria Digital Marketing. The top 20 valuable Facebook statistics, < https://zephoria.com/top-15-valuable-facebook-statistics/ > ; July 2016.

[6] Cloud networking: scaling out data center networks, Arista whitepaper, < https://www.arista.com/assets/data/pdf/Whitepapers/Cloud_Networking__Scaling_Out_Data_Center_Networks.pdf > ; 2016.

[7] Teach yourself fat-tree design in 60 minutes, < http://clusterdesign.org/fat-trees/ > .

[8] High-density 25/100 gigabit ethernet StrataXGS Tomahawk ethernet switch series, < https://www.broadcom.com/products/Switching/Data-Center/BCM56960-Series > .

[9] Calient uses its optical switch to boost data centre efficiency, < http://www.gazettabyte.com/home/2015/4/17/calient-uses-its-optical-switch-to-boost-data-centre-efficie.html > ; April 17, 2015.

[10] Rockley demos a silicon photonics switch prototype. Gazettabyte, < http://www.gazettabyte.com/home/2015/9/27/rockley-demos-a-silicon-photonics-switch-prototype.html > ; September 27, 2015.

[11] Intel's Rack scale architecture, < http://www.intel.com/newsroom/kits/atom/c2000/pdfs/Architecting_for_Hyperscale_DC_Efficiency.pdf > .

[12] The optical backplane is finally here. Will this change everything? LightCounting market research, research note; July 20, 2016.

[13] Chip shot: Intel and corning demonstrate record-breaking reach of optical fiber, < https://newsroom.intel.com/chip-shots/chip-shot-intel-and-corning-demonstrate-record-breaking-reach-of-optical-fiber/ > .

[14] MXC Spring 2014 adopters—Intel, < http://www.intel.com/content/dam/www/public/us/en/ documents/intel-research/mxc-spring-2014-adopters.pdf >.

[15] Intel delays part for high-speed silicon photonic networking, < http://www.computerworld. com/article/2879077/servers/intel-delays-part-for-high-speed-silicon-photonic-networking. html >.

[16] Light at the end of Intel's Silicon Photonics: 100 Gbps network tech finally shipping, Sorta, < http://www.theregister.co.uk/2016/08/17/intel_silicon_photonics/ >.

[17] Ericsson introduces a hyperscale cloud solution, < https://www.ericsson.com/res/docs/2015/ intel-ericsson-solution-brief.pdf >.

[18] Fibre-to-the-NPU: optics reshapes the IP core router. Gazettabyte, < http://www.gazetta-byte.com/home/2013/3/12/fibre-to-the-npu-optics-reshapes-the-ip-core-router.html >; March 12, 2013.

[19] Ethernet opens door to PAM4. Mark Nowell blog, < http://www.eetimes.com/author.asp? section_id = 36&doc_id = 1326889 >; June 17, 2015.

[20] High speed silicon photonics links over multimode fiber, Scott R. Brickham presentation, slide 8, presented at photonics for disaggregated data centers workshop, Los Angeles, CA; March 22, 2015.

[21] Mellanox readies first 200 Gb/s silicon photonics devices. HPCwire, < https://www.hpcwire. com/2016/03/22/mellanox-reveals-200-gbps-path/ >; March 22, 2016.

[22] What's the difference between NRZ and PAM? Electronicdesign, < http://electronicdesign. com/communications/what-s-difference-between-nrz-and-pam >; June 8, 2015.

[23] The state of ethernet optics. In: Presented at OFC, < http://www.ethernetalliance.org/wp-content/uploads/2016/01/The-State-Of-Ethernet-Optics-Final.pdf >; March 23, 2016.

[24] Consortium for on-board optics, < http://cobo.azurewebsites.net/ >.

[25] 2D vs. 2.5D vs. 3D ICs 101. EE times, < http://www.eetimes.com/document.asp?doc_-id = 1279540 >; April 8, 2012.

[26] FPGAs with 56-gigabit transceivers set for 2017. Gazettabyte, < http://www.gazettabyte. com/home/2016/6/29/fpgas-with-56-gigabit-transceivers-set-for-2017.html >; June 29, 2016.

[27] Start-up Sicoya targets chip-to-chip interfaces. Gazettabyte, < http://www.gazettabyte.com/ home/2016/2/10/start-up-sicoya-targets-chip-to-chip-interfaces.html >; February 10, 2016.

[28] Sun, et al. Single-chip microprocessor that communicates directly using light.. Nat Int Wkly J Sci December 2015;528(7583):534−8.

CHAPTER 8

The Likely Course of Silicon Photonics

A strategic inflection point is a time in the life of business when its fundamentals are about to change. That change can mean an opportunity to rise to new heights. But it may just as likely signal the beginning of the end.

Andrew S. Grove, former Intel CEO [1]

The world is about to change, and I don't think people have quite figured that out.

Professor John Bowers on silicon photonics [2]

What prediction really comes down to is studying history, looking hard at our current moment... and then — guessing.

Kim Stanley Robinson, science fiction writer [3]

8.1 LOOKING BACK TO SEE AHEAD

This final chapter of the book reflects on the likely course of silicon photonics near term and offers a perspective on longer-term possibilities.

A look back 20 years demonstrates how quickly technologies can emerge. It also reveals the perils of prediction.

Back in 1995, the first version of Microsoft's Internet Explorer web browser was launched; data and messaging services using GSM—the second-generation cellular wireless standard—began; and you could access the Internet with a download speed of several tens of kilobits a second using a dial-up modem. Amazon was a 1-year-old company, while cloud computing, smartphones, Big Data, YouTube, and Google were all to come.

Looking forward, there is a good reason to believe that the pace of change will be even quicker in the next two decades. Unlike in 1995, a

Silicon Photonics. DOI: http://dx.doi.org/10.1016/B978-0-12-802975-6.00008-9
Copyright © 2017 Daryl Inniss and Roy Rubenstein. Published by Elsevier Inc. All rights reserved.

mature wireless and wireline networking infrastructure is now in place, and many things including innovation are happening faster [4].

Accordingly, silicon photonics is emerging into a technological future that will be strikingly different from what we recognize today. A faster pace of change will benefit silicon photonics and inevitably dictate its future course. Hence, it is right to be cautious, but there is value in forecasting, extrapolating current trends over differing time spans to predict silicon photonics' impact.

8.2 THE MARKET OPPORTUNITIES FOR SILICON PHOTONICS: THE PRESENT TO 2026

In a 2016 report [5] LightCounting Market Research analyzed integrated optical component trends and the role of silicon photonics compared to established optical component technologies such as indium phosphide and gallium arsenide.

LightCounting's report included several notable findings. One finding is that only 1 in 40 optical components is an integrated design. Here, optical components refer to transceivers and lasers but not components such as optical amplifiers and optical switches. This means that in 2016, less than 3% of all the optical components sold in the telecom and datacom markets were integrated—a surprisingly small fraction.

Yet while integrated components account for a small fraction of total volumes, they generate one-third of the total annual optical component market revenues. Integrated devices are valuable designs due to their relatively high average selling price.

Looking ahead, LightCounting concludes the market for silicon photonics devices will grow threefold to $1 billion in 2021. By then, 1 in 10 optical components will be integrated, and will account for 60% of global revenues. Total sales of integrated devices, including indium phosphide and gallium arsenide products, are projected to reach $5 billion by 2021.

LightCounting concludes that the optical component market impact of silicon photonics will not be significant in the near term: 2016–21. Still, a tripling of market value to $1 billion represents significant growth. Such revenue growth is welcome but will be shared among many silicon photonics companies.

The 2021 $1 billion market also does not include the use of silicon photonics by companies within their own systems, players such as Huawei, Mellanox, Cisco, Ciena and Juniper using the technology in their servers, transceivers, or switches. Nor does it include silicon photonics products from emerging start-ups such as Ayar Labs, Sicoya, and Rockley Photonics.

Such examples are not counted as direct optical component sales but these developments will generate revenues for the parent companies and will advance silicon photonics. They also represent examples of the core benefit of silicon photonics, working alongside electronics as part of system designs that give companies a technology edge.

8.2.1 Near-Term Opportunities: 2016–21

Fig. 8.1 shows the likely timescales of emerging markets for silicon photonics as well as important developments. As shown, telecom and datacom are markets where silicon photonics is already playing a role, and this is the obvious main opportunity for the next 5 years.

These opportunities include 100-Gb transceivers from mid-reach links in the data center and 100-Gb and faster modules for coherent optical transport. Mid-reach optics spans 0.5–2 km and is served currently by such module designs as PSM4 and the CWDM4 and CLR4. There are also 100-, 200-, and 400-Gb coherent applications using the CFP and CFP2-ACO designs.

It is also likely that multiwavelength terabit-plus coherent photonic integrated circuits will appear, as indicated by the emerging CFP8-ACO pluggable module [6].

Also shown are Microsoft's Madison module requirements. Microsoft has data centers made up of several buildings distributed on a campus that are separated by distances of 2 km. It also has buildings making up a data center spread as far as 70 km apart. The Madison modules are the optical components industry's answer to Microsoft's demand for optics outside standard multisource agreement initiatives—an example of an Internet content provider driving new optical developments.

The first Madison QSFP28 module will support 100 Gb using just two wavelengths, each 25 Gb, coupled with four-level pulse-amplitude

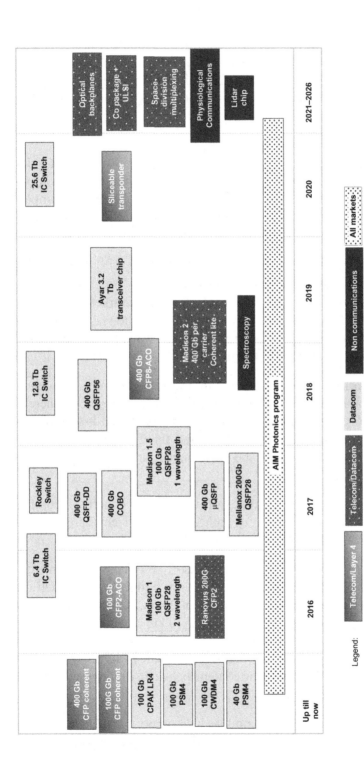

Figure 8.1 Opportunities for silicon photonics and application timings.

modulation (PAM4) signaling. Microsoft is working with semiconductor supplier Inphi to develop the module. Microsoft is also talking to interested optical module makers to develop a follow-on, known as Madison 1.5, to create a 100-Gb QSFP28 design using one wavelength only based on 50-GBd signaling and PAM4. The specification is to achieve a total bandwidth of between 6.4 and 7.2 Tb down a fiber.

Lastly, Microsoft is also interested in a tailored coherent optics solution that will allow up to 38 Tb transmission. Madison 2.0 is designated to be implemented using the Consortium of On-Board Optics (COBO) form factor (see Sections 5.5.1 and 7.5.4).

One multisource agreement, the OpenOptics wavelength division multiplexing design cofounded by Ranovus and Mellanox, is being aimed at higher-than-100-Gb links within the data center [7]. New higher-capacity pluggable form factors will also be launched during this period, such as the QSFP-DD, the μQSFP, and the QSFP56.

Data center networking will also drive new Ethernet speed standards such as 25-, 50-, 200-, and 400-Gb Ethernet. The simplest way to implement 50-Gb signaling will be to use two 25-Gb lanes but that will quickly be replaced with more elegant 50-Gb single lanes. A 50-Gb lane—25-GBd optics and PAM4 modulation—will enable these new-speed Ethernet configurations. And the combination of 50-GBd optics and PAM4 will enable 100-Gb single lanes. Mellanox is one company that will use 50-Gb nonreturn-to-zero signaling and is confident that the approach can be extended to 100-Gb using its silicon photonics technology. But at 100-Gb, copackaged optics will be needed.

Midboard optics designs using COBO are also to be expected, with first hardware supporting 400 and 800 Gb/s projected toward the end of 2017. Midboard optics will become a more pressing need as switch chip companies announce 6.4-, 12.8-, and 25.6-Tb switch silicon.

New system architectures using silicon photonics will also be deployed in the near term, such as Rockley's switch architecture for the data center, and chip-to-chip optical designs. Other innovations will be developed by start-ups that are still in stealth mode, and silicon photonics will also be adopted by systems vendors in their own equipment, developments that are not always announced.

In turn, it is possible that high-capacity sliceable transponders will emerge for long-distance optical transmission toward the end of the near term.

Thus the number and diversity of silicon photonics products will grow as we move into the 2020s.

While this book has focused on emerging silicon photonic opportunities for datacom and telecom, spectroscopic products will also benefit from photonic integration on a CMOS platform. The new Fourier transform infrared spectrometer being developed by Lumux Technology is one example [8].

The sensor is a potential high-volume, low-cost product for applications like environmental monitoring in the oil and gas industries. The company's goal is a consumer spectrometer that works with personal mobile devices.

8.2.2 Mid-Term Opportunities: 2021−26

Looking out beyond 2021 to the following 5 years—also shown in Fig. 8.1—the opportunities become more speculative. But clearly the next Ethernet speed grade of 400 Gb and its successors can be expected. Beyond 400-Gb Ethernet could be 800-Gb or higher multiples of 400: 1.2-Tb Ethernet or even 1.6-Tb Ethernet. Other intermediate Ethernet rates should also be expected.

Midboard optics and optical backplanes will become more commonplace [9], as well as short-link silicon photonics−based connections served currently using VCSELs. In the mid-term we will also likely see examples of optics copackaged with ultralarge-scale integrated circuits such as 25.6-Tb switch chips.

Telecom markets for silicon photonics such as data center interconnect—coherent and noncoherent—as well as mobile applications driven by the adoption of mobile 5G technology should also add to volumes.

And then there are the emerging markets for silicon photonics such as sensing for the Internet of Things, medical devices, and light detection and ranging (LIDAR) [10]. Nokia Bell Labs has long been active in the area of silicon photonics and one of its programs concerns physiological communications and the development of sensors that integrate optics and wireless technologies for monitoring and operating inside the human body. This is one example of sensor devices that could drive silicon photonics into mass-market applications.

And while there is still a question mark regarding space-division multiplexing, it could represent a significant telecom opportunity for silicon photonics longer term.

Looking at the next decade, several conclusions can be drawn:

• The data center remains the most important market opportunity for silicon photonics.
• The number of applications using silicon photonics is growing over time
• A killer application for silicon photonics has yet to emerge.

8.3 THE GREAT CULTURAL DIVIDE

All the chapters up till now have focused on technology. How technology takes hold in the marketplace, semiconductor technologies, photonics technologies, and technologies used in systems. Such a tech-centric focus is necessary to assess the significance of silicon photonics and identify where it will play a role.

Yet the biggest challenge facing silicon photonics does not concern the technology. That is not to downplay the technical and manufacturing obstacles that remain. It is just that with investment and time, such challenges are being overcome. Silicon photonics is gaining momentum and the challenges being addressed are engineering ones, not issues awaiting a fundamental technological breakthrough.

Silicon photonics' biggest challenge is, in fact, people. Or, more accurately, the cultural divide between the chip and the photonics communities.

The optical component industry has a tradition of making relatively small numbers of custom devices. And when the industry has embraced standardization such as pluggable optical components, competition has been fierce, and the ability for any one company to make differentiated products has been limited. The advent of the data center market has brought shorter product cycles and the promise of volumes but for now the optical component industry is not prepared for that.

Silicon photonics brings high-volume manufacturing and the promise of integration. But the optical components industry has limited opportunity for system integration and significant volumes for such products do not exist.

In contrast, the chip industry is a master of high-volume manufacturing, it has widely embraced and benefited from standardization, and unlike the optical component industry, it can use billions of transistors to produce differentiated products. The chip vendors also have more scope to develop systems, albeit systems at the chip level.

But chip engineers are skilled in electronics design and have little knowledge of photonics and are unlikely to start changing how they design chips. This explains why optics in the data center is largely confined to pluggable optics on the edge of systems and to cabling, even though the Internet content providers want optics brought inside systems.

In August 2016 Juniper Networks announced its intention to acquire US silicon photonics start-up Aurrion [11]. Juniper joins several systems vendor that have made acquisitions to bring silicon photonics technology in-house. Such moves by the equipment makers are an acknowledgment that as photonics moves closer to the silicon and away from a system's faceplate, silicon photonics is becoming strategically important.

So will the telecom and datacom system vendors, that design their own chips and now have in-house photonics expertise, be the ones to perform a mashup and bridge this cultural divide? The answer is most likely yes, or at least they will be trailblazers, but only for their benefit.

A consequence of these acquisitions is that the silicon photonics technology becomes the property of the systems houses and is lost to the market at large. For the wider community, Aurrion represents yet another silicon photonics start-up whose technology has been removed for use by the wider marketplace.

This is what AIM Photonics—included in Fig. 8.1—is looking to address. AIM Photonics is a US public–private partnership that is developing technology for integrated photonics. In particular, AIM is looking to advance the manufacturing of silicon photonics, making the technology available to small-to-medium-sized businesses and entrepreneurial companies.

Silicon photonics luminary, Professor Lionel Kimerling, sees AIM Photonics as bringing the manufacturing discipline of the electronics industry to photonics. AIM will also allow chip designers to start exploring the design benefits of silicon photonics. AIM, e.g., will be offering a system-in-package service by the end of 2018. AIM Photonics is a 5-year project to be completed in 2020 by when it hopes to be self-sustaining [12].

This explains why Professor Kimerling is spending much of his time putting together educational material to help attract individuals to pursue a career in silicon photonics. Much of the technology is in place, he says, what is required is to make it accessible to people. "Once we get it accessible and people start to experiment and design, good things are going to happen," says Kimerling.

AIM Photonics has the potential to help bring new silicon photonics players and their products to market but it alone will not bridge the cultural divide. What will is when electronics and photonics replaces Moore's law as the basis for system scaling. The data center and its requirements are already showing that this is starting to happen.

In Chapter 1, Silicon Photonics: Disruptive and Ready for Prime Time, we asked if silicon photonics is disruptive. Any emerging technology that promises to continue scaling post-Moore's law suggests it is disruptive. This new scaling force will also cause silicon photonics to cross over to the chip industry. Silicon photonics driven by the chip industry with also disrupt the existing optical industry, as is now explained.

8.4 THE CHIP INDUSTRY WILL OWN PHOTONICS

The need to continually scale systems will mean that the chip industry will sooner or later embrace silicon photonics. Silicon photonics will then move from a technology being developed by the optical industry to one being driven by the much larger chip industry. Or as silicon photonics luminary and Aurrion cofounder Professor John Bowers puts it, photonics will transfer to silicon [2].

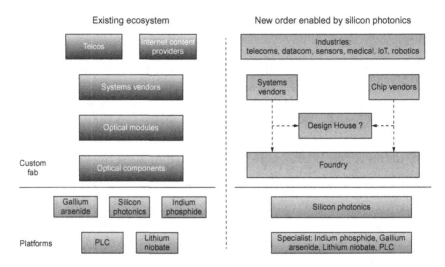

Figure 8.2 The optical industry now and in the future.

This will significantly change the optical industry which, today, has a hierarchical structure (see Fig. 8.2, *left*). It is not clear whether silicon photonics design houses will emerge, as ASIC design houses did for companies that wanted custom chips. But ultimately the foundries and large chip companies with their own fabrication plants will come to own silicon photonics.

Already there are chip foundries making silicon photonics chips, while leading chip companies such as Intel and STMicroelectronics are actively developing silicon photonics chips in their fabs. Both Intel and STMicroelectronics offer 100-Gb PSM4 module designs. But their motivation to embracing silicon photonics was not to become optical transceiver players. Meanwhile, the systems vendors with their silicon photonics acquisitions are already moving toward this new ecosystem.

Silicon photonics will become the optical integration platform, just as CMOS became the platform for the bulk of the chip industry. And similarly, while not all electronic chips are designed in CMOS, silicon photonics will dominate volumes but indium phosphide, gallium arsenide, lithium niobate, and planar lightwave circuits will continue to be made. But they will become increasingly specialist; silicon photonics will become the platform for optics.

By 2035, people will not think in terms of CMOS and silicon photonics. Design tools will guide developers as to what is best done electrically and what is best done optically. Such tools will span systems: how chips will be copackaged, how system-in-package devices will be integrated on a board and how such boards will be combined to make systems. And of course system design will cover new applications. In-body sensor devices will have very different systems requirements to today's telecom and datacom platforms.

Specialist functions and materials will also emerge in the next 20 years. There will be other forms of computation such as optical alongside digital, and more materials such as graphene will be added to the mix. These system elements will need to be stitched together and communicate to create more complex and varied designs. So while photonics and electronics may be coming together, the form factors and the technologies will likely become more varied.

It may be premature to talk about silicon photonics' legacy when the industry is still openly questioning its merits and significance. But our expectation is that, as its name implies, silicon photonics will be viewed as the technology that bridged two distinct industries: photonics and semiconductors.

Indeed, silicon photonics will come to be seen as nothing less than a strategic inflection point of the Information Age.

Key Takeaways

- The near-term opportunities for silicon photonics will continue to be datacom and telecom, with new applications such as sensors, LIDAR, and microwave optics emerging over the next decade.
- The biggest challenge facing silicon photonics comes down to people and in particular the cultural divide between the chip and the photonics communities.
- Silicon photonics will help chip and systems performance to continue to scale post-Moore's law.
- Silicon photonics is a disruptive technology. It will also disrupt the existing optical supply chain.
- The chip industry will ultimately drive optics, with silicon photonics becoming an integral part of its offerings. Indeed, at some point in the future, the very term silicon photonics as a differentiating nomenclature will cease to be relevant.

REFERENCES

[1] Grove AS. Only the paranoid survive. New York: Random House; 1996.

[2] Heterogeneous integration comes of age. Gazettabyte, <http://www.gazettabyte.com/home/2016/8/28/heterogeneous-integration-comes-of-age.html>; August 28, 2016.

[3] The great unknown. Sci Am, 2016;315(3):75−9.

[4] Colvile R. The great acceleration: how the world is getting faster, faster. Bloomsbury; 2016.

[5] Is silicon photonics a disruptive technology? Market opportunity for optical integration technologies, LightCounting Market Research report; January 2016.

[6] OIF starts works on a terabit-plus CFP8-ACO module. Gazettabyte, <http://www.gazettabyte.com/home/2016/7/24/oif-starts-work-on-a-terabit-plus-cfp8-aco-module.html>; July 24, 2016.

[7] The OpenOptics MSA, <http://www.openopticsmsa.org>.

[8] Lumux Technology Corp. company website, <http://www.luxmux.com>.

[9] The optical backplane is finally here. Will this change everything? LightCounting Market Research, research note; July 20, 2016.

[10] MIT and DARPA pack lidar sensor onto single chip. IEEE Spectrum, <http://spectrum.ieee.org/tech-talk/semiconductors/optoelectronics/mit-lidar-on-a-chip>; August 4, 2016.

[11] Juniper networks to acquire Aurrion for $165 million. Gazettabyte, <http://www.gazettabyte.com/home/2016/8/8/juniper-networks-to-acquire-aurrion-for-165-million.html>; August 8, 2016.

[12] Making integrated optics available to all. Optical Connections, Issue 7, Q3 2016, p. 16.

Optical Communications Primer

A1.1 OPTICAL LINKS

Communication systems involve the sending of information—typically electrical signals carrying digital data—between two points. Even for optical communication, which uses fiber to carry information in the form of light, electrical signals are first converted to optical ones before transmission, and converted back from optical to electrical at the receiver.

Communication is either unidirectional, known as simplex, or bidirectional (duplex). If both directions operate simultaneously, it is known as full-duplex [1].

For simplex communications, a transmitter is used at one end of the link and a receiver at the other—one to send data and one to receive. When bidirectional communication is involved, a receiver and transmitter pair—a transceiver—is required at each end.

The translation between electrical and optical signals for optical communication adds complexity and cost. So why do it? Because there are telling advantages when communicating optically. The bandwidth—the information-carrying capacity of the optical medium—is far greater in the optical domain than in the electrical domain, as is the reach.

Typically the receiver and transmitter are copackaged in an optical transceiver, also referred to as an optical module.

The main ways data is sent optically are shown in Fig. A1.1:

1. *Single-channel transmission*: Data transmission using a single wavelength on a single fiber.
2. *Parallel fiber*: Multiple fibers in the form of a ribbon cable are used, with each fiber performing single-channel transmission as in scheme 1. The parallel transmission's capacity is the number of fibers multiplied by channel data rate.

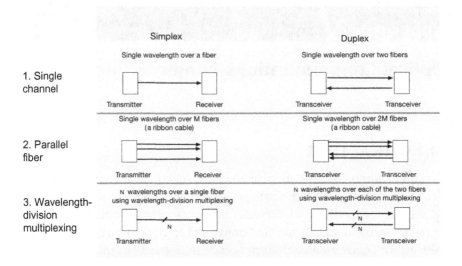

Figure A1.1 The main three approaches used for optical communications.

3. *Wavelength-division multiplexing*: A technique to send multiple wavelengths over a fiber. This is another form of multichannel transmission but it uses one fiber, with each channel being carried using a distinct wavelength of the fiber's spectrum. Coarse wavelength-division multiplexing (CWDM) uses 8 or 16 wavelengths, each relatively wide apart. Dense wavelength-division multiplexing (DWDM) refers to higher channel counts where the wavelengths are packed more closely across the fiber's spectrum.

For duplex communications, the total fiber count is double the schemes above. One twist on these schemes is the bidirectional approach that scales capacity using wavelengths to avoid having to use multiple parallel fibers. The technique sends data in both directions on one fiber using a different wavelength for each.

These main three techniques help designers choose the degree of complexity needed to meet the optical performance requirements of a given application.

A single channel is the simplest and cheapest scheme. There is one laser at the transmitter and one photodetector at the receiver. But only so much data can be sent across the channel. The current highest-speed single-channel rate is 25 Gb/s, although a 40-Gb/s 2 km serial standard has been deployed commercially.

For short-reach applications, the transmitter's laser is directly modulated. Here the laser is turned on and off to encode data so that a separate modulator is not required. But a modulator is used when higher optical performance and longer link distances are needed.

The modulator's role, at its simplest, is like a card waved in front of the light source, to pass or block light as required to encode the data being sent. Naturally, adding a modulator adds cost to the transceiver.

The parallel, multiple-lane approaches—schemes 2 and 3 above—are used when the data to be sent exceeds the capacity of a single channel.

One approach is to use single channels in parallel in the form of a ribbon cable. The link is composed of a transceiver at each end connected with a ribbon cable. The ribbon fiber, with each strand carrying a wavelength, is attractive for shorter links. Each channel is relatively simple, but using multiple fibers adds cost especially as the link distance increases. Ribbon cable also adds wiring complexity that needs to be managed and routed within a data center.

One such example of a parallel link is the IEEE 100GBase-SR4 standard, as shown in Fig. A1.2. A 100-m, 100-Gb IEEE standard for the data center, the link uses four multimode fibers, each transmitting 25 Gb, and four fibers each receiving 25 Gb.

The Parallel Single Mode 4 - PSM4 - multisource agreement is another example using this approach.

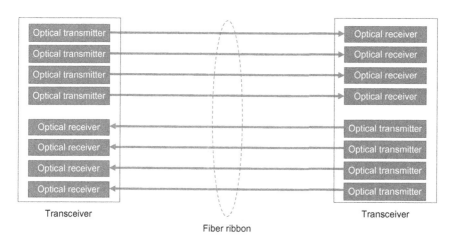

Figure A1.2 A parallel optical link illustrating four transmit and receive lanes.

Another example is the emerging short-reach 400-Gb Ethernet interface, dubbed 400GE-SR16, which uses sixteen 25-Gb channels in each direction. Accordingly, the 400-Gb short-reach interface will have 32 multimode fibers overall, an example of how channel count rises with data rate.

The second parallel approach is wavelength-division multiplexing. Here, channel parallelism is achieved using wavelengths across the fiber's spectrum, commonly single-mode fiber. To send wavelengths across a fiber, multiple pairs of lasers and receivers are used as well as another optical function pair—a multiplexer and demultiplexer.

The multiplexer acts as an on-ramp, placing each wavelength onto the fiber, while the demultiplexer separates each wavelength at the destination. The difference from ribbon fiber is that a more complex transceiver is required at both ends, but for wavelength-division multiplexing only one fiber is used in each direction for full-duplex communication.

One example of wavelength-division multiplexing scheme is the 2-km CWDM4 multisource agreement, another 100-Gb interface for the data center. The CWDM4 transceiver uses four lasers and four photodetectors as well as a multiplexer and demultiplexer, with each wavelength carrying 25 Gb. Fig. A1.3 shows the optical functions of the CWDM4 transceiver.

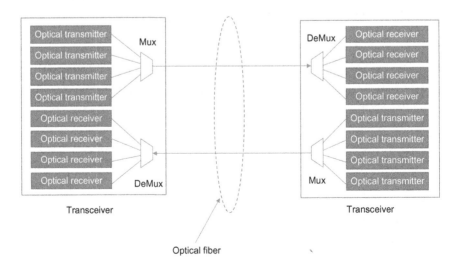

Figure A1.3 A CWDM4 transceiver optical link using wavelength-division multiplexing.

Wavelength-division multiplexing is the bedrock of long-distance optical transmission. A complex transceiver design is needed to send and receive each wavelength hundreds or thousands of kilometers, each one typically carrying a 100-Gb signal. Up to 96 such wavelengths can be packed on a single-mode fiber.

A1.2 OPTICAL COMPONENT TECHNOLOGIES

Optical designers must consider the following performance metrics when deciding which optical component technology to use:

- *Bandwidth*: The capacity of the link, which determines how much information or data, measured in bits per second, can be sent over a fiber or waveguide.
- *Distance*: The link's reach. Sending data across an optical link requires a certain laser power and a certain photodetector sensitivity to detect the light at the fiber's end. Clearly, sending data between switches up to 100 m apart is very different from sending data over a pan-Pacific submarine link. The channel affects the transmission—the signal is attenuated in transit and suffers distortion. Such impairments limit the reach beyond which the data cannot be recovered.
- *Cost*: Here, cost refers to the electronics and optics at each end of the link, or the total link including the fiber. The cost can be for discrete components—optical components and chips on a line card—or for components integrated in a "pluggable" optical module that slots into data center equipment. The cost of the fiber as well as the transceivers may also be considered, especially if the use of multi-stranded fiber ribbon cable is an option.

A1.2.1 Short-Reach Links

Copper cabling is used for 10-Gb high-speed links up to 10 m, typically on twinaxial (twinax) cable. Copper cabling can also support 25 Gb up to 5 m. Such an electrical link is also the cheapest; no translations are needed between the electrical and optical domains.

Optical links are used for longer reaches and for higher transmission speeds. The shortest distances for which optical interconnect is used—10 m to several hundred meters—are the most cost-sensitive and require the lowest-cost transceiver technologies. Such transceivers use directly modulated vertical-cavity surface-emitting lasers (VCSELs).

VCSELs are made using gallium arsenide, a III-V compound, and operate at 850 nm over multimode fiber. Inexpensive photodiodes at the receiver are also used. Multimode fiber is more expensive than single-mode, but the total cost of the link is less because of the inexpensive VCSEL-based transceivers at each end.

For 40- and 100-Gb links, four multimode fibers are used in parallel, each lane consisting of an 850-nm wavelength operating at 10 or 25 Gb, respectively. Commercial VCSELs can work at rates as high as 28 Gb; the 32-Gb Fibre Channel storage interface standard, e.g., operates at 28.05 Gb. Finisar has demonstrated VCSELs working at 56 Gb/s [2].

For 40-Gb transmission, four fibers each at 10 Gb are used to transmit and four to receive, an IEEE standard known as 40GBase-SR4 (SR standing for short reach). Similarly, for 100-Gb links, the four fibers each transmit 25 Gb of data per second and four fibers receive— the 100GBase-SR4 standard.

A1.2.2 Mid-Reach Optics

Single-mode fiber is used for distances beyond a few hundred meters and, for the purposes of this discussion, shorter than 10 km, referred to as mid-reach. Two 100-Gb solutions have been developed: the 500-m PSM4 and the 2-km CWDM4. There is also the CLR4 which is similar in design to the CWDM4 but has a more stringent optical specification. Some optical module vendors offer a pluggable module that supports the CWDM4 and the CLR4 solutions [3].

The PSM4 uses four single-mode fibers for transmission and four to receive, as mentioned. Here, each laser used is not a gallium arsenide VCSEL but a directly modulated edge emitter laser at 1300 nm, implemented using indium phosphide to meet the more demanding reach requirements.

In contrast, the CWDM4 uses four wavelengths across one fiber, using spectrum around 1300 nm. Since wavelength-division multiplexing is used, each of the four channels has a unique wavelength. Mid-reach interfaces are cost-sensitive given their use for the data center. The CWDM4 is a coarse wavelength-division multiplexed design, as the name implies, with a generous 20 nm between wavelengths. This is wide enough to tolerate some wavelength drift due to changes in the operating temperature of the lasers.

A1.2.3 Long-Reach Optics

To support transmission to 10 km, known as long reach, the four wavelengths are bunched closer together: 2−4 nm apart. The wavelengths are close enough that temperature control is needed to ensure each laser does not drift away from its wavelength; a thermoelectric cooler is used to keep the laser's temperature stable, adding to the transceiver cost. Otherwise, the architecture and link are similar to the CWDM4. The long-reach standard—an example of local area network wavelength-division multiplexing (LAN WDM)—is called 100GBase-LR4.

For distances beyond 10 km, e.g., between sites, the IEEE standardized the 100GBase-ER4 that supports 40-km 100-Gb transmissions. The optics and the link are similar to those of the LR4, using four LAN wavelength-division multiplexed wavelengths in the 1300-nm band. Here, an electroabsorption modulated laser is used; the laser and modulator functions are separated on the chip, and the laser has fewer frequency variations than a directly modulated laser. The resulting laser is larger, consumes more power, and is more expensive.

Customers want to use the same transceiver form factors for longer distances that they use for shorter-reach ones, presenting significant technical challenges for optical module designers.

A1.2.4 Long-Distance Optics

Tunable lasers are used for Layer 4 metro and long-haul distances. Here, the laser can be set at a particular wavelength around the long-distance 1550-nm band. External modulators, typically in lithium niobate, are used. Such transmitters support dense wavelength-division multiplexing and achieve transmission distances of thousands of kilometers.

In summary, the optics that are adopted for a transceiver depend on the application:

- Multimode lasers and receivers operate at 850 nm for short-reach applications—10 m to a few hundred meters.
- Directly modulated lasers and electroabsorption modulated lasers operate at 1310 nm using single-mode fiber, achieving several tens of kilometers.
- Tunable lasers and external modulators operate at wavelengths around the 1550-nm wavelength for transmissions of up to thousands of kilometers.

A1.3 ATTENUATION CHARACTERISTICS OF FIBER

We mention different operating wavelengths in fiber: 850 nm for multi-mode fiber, and 1300 and 1550 nm for single-mode fiber. These are used with good reason: they are wavelength bands for the optical fiber where signal attenuation, measured in dB/km, is low and suits the transmission distances.

The attenuation curve in Fig. A1.4 makes clear why 1300 and 1550 nm are used for medium- and long-distance transmission since the loss is a fraction of a decibel per kilometer. Optical amplification and dispersion are other factors contributing to the transmission windows. Although the lowest loss and dispersion is at 1300 nm, the 1550-nm window emerged as a better wavelength for WDM transmission. One counterintuitive reason for this is that zero dispersion, which for a germanosilicate fiber is near 1300 nm, leads to nonlinear transmission impairments. Consequently, optical amplifiers and wavelength-division multiplexing were developed in the 1550-nm window.

Most long-distance transmission is between 1525 and 1565 nm in what is called the conventional or C band. The broad-based development and commercialization of erbium amplifiers is largely responsible for the C band's popularity.

Erbium also amplifies between 1570 and 1610 nm, facilitating long-distance transmission in this long or L band region. As C and L band

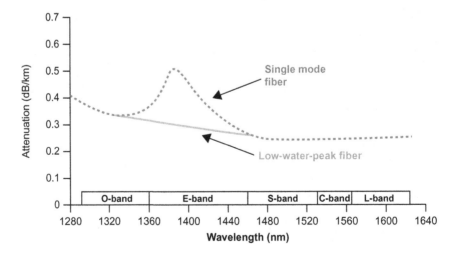

Figure A1.4 Optical attenuation curve of a germanosilicate fiber.

amplifier designs are different, two amplifiers are needed if transmission in both bands is desired. In Chapter 5, Metro and Long-Haul Network Growth Demands Exponential Progress, we discuss the C and L bands.

Operating at 850 nm, the attenuation is higher but the distances are substantially shorter. In fact, distance is limited by modal dispersion, the temporal spreading of the transmitted modes due to different propagation velocities.

Note that around 1400 nm, there is an absorption peak due to the effect of traces of water in the glass fiber. Fabrication techniques can be used to lower this absorption below 0.5 dB/km to expand the transmission window or fiber spectrum over which data can be sent.

A1.4 OPTICAL MODULES

Optical modules arose with the advent of 10-Gb/s transmission prior to the optical boom of 1999–2000.

At that time, equipment vendors such as Nortel and Lucent Technologies had their own optical component divisions to make and design optical interfaces for their systems. But around 2000, the then-leading telecom equipment makers divested their component divisions, ceding their component know-how to optical module and optical component companies. The equipment makers realized that their focus and cash could be better spent elsewhere, and that the influx of module-making companies during the optical boom would drive down the cost of interfaces.

This has been the business model for the past 15 years. Transceiver companies were able to aggregate volumes by selling to multiple equipment vendors and this, coupled with the fierce competition, has helped drive down costs.

For 10-Gb long-haul transmission, the first designs to market were line cards containing discrete state-of-the-art optical and chip components. Ten gigabit transmission was also the last speed to use simple on-off keying. This modulation scheme was also used for 40-Gb links that followed, but the move to 40 Gb also marked the introduction of more advanced modulation schemes based on phase that are now standard for 100-Gb and greater link speeds.

The first 10-Gb optical modules developed by third-party optical module companies were not alternatives for the highest performance 10-Gb links made by the equipment makers, but they achieved long-distance dense wavelength-division multiplexing–based transmission and offered a buying alternative for suitable links.

The first such modules were 5-by-7-in. 300-pin MSA modules. These were necessarily large to fit in all the components needed for long-haul performance. The 300-pin MSA supported 10-Gb transmission and was quickly followed by a 40-Gb MSA [4].

The market realized that Layer 3 applications provided the highest volumes and that small size was paramount for such applications. What followed was a string of announcements of ever-smaller form factors that ultimately led to the SFP + (see Table A1.1).

The downside of such choice was that optical module makers had to choose which form factors to back at a time when it was unclear what end users would want.

A1.4.1 The Miniaturization of Modules
The SFP + form factor has become the de facto module of choice for high-density designs—used for 48-port Ethernet switches inside the data center, e.g., this yields a total capacity of 480 Gb.

But it took more than a decade to move 10-Gb line-side technology from a discrete line card to fit into the SFP + module. This may sound like a long time, but it is a remarkable achievement in terms of cramming ever-more-complex designs into the same form factor. There is now a suite of 10-Gb products in the SFP + form factor, from 10-Gb short-reach interfaces to the tunable SFP + , where the tunable laser

Table A1.1 Optical Module Form Factors and Their Relative Sizes

10G Form Factor	MSA Published	Size (Volume Relative to SFP +)
300 pin	2001	x21.2
XENPAK	2001	x11.6
XPAK	2002	x7.9
X2	2003	x6.9
XFP	2003	x1.7
SFP +	2006	x1

supports dense wavelength-division multiplexing transmission over several hundreds of kilometers.

The SFP + is the most common pluggable optical module. Having an optical module that can be plugged into and detached from telecom and datacom equipment brings several benefits:

- Operators can replace the module should a failure occur without having to replace the complete line card, which would disrupt other links.
- The same form factor can support different interfaces, including future ones with enhanced performance. This enables easy upgrades once a platform is deployed.
- The operator of equipment can install interfaces as needed, saving up-front costs.
- Purchasers can choose among transceivers from multiple vendors based on cost and performance.

Today, the important pluggable form factors besides the SFP + include the quad-channel version of the SFP +, known as the QSFP +, and the 28-Gb versions of the two—the SFP28 and the QSFP28, the latter of which is the 100-Gb form factor of choice in the data center.

Another important family of optical modules is the CFP MSA family [5], which includes the CFP, CFP2, CFP4, and CFP8 (see Fig. A1.5). There is a deliberate reduction in size from the CFP to the CFP4 to ensure ever-increasing interface densities on equipment. The CFP8 is approximately the size of the CFP2 and is being earmarked for 400-Gb Ethernet and, for line-side optics, the CFP8-ACO. A bigger size is needed to accommodate the greater number of channels 400-Gb requires and the greater heat such designs will generate.

CFP   CFP2   CFP4   CFP8

Figure A1.5 The CFP MSA family members. Finisar.

REFERENCES

[1] Stremler FG. Introduction to communication systems. 3rd ed. Reading, Massachusetts: Addison-Wesley Publishing Company; 1990.

[2] Achieving 56-gigabit VCSELs. Gazettabyte, <http://www.gazettabyte.com/home/2013/2/1/achieving-56-gigabit-vcsels.html>; February 1, 2013.

[3] Intel's 100-gigabit silicon photonics move. Gazettabyte, <http://www.gazettabyte.com/home/2016/8/21/intels-100-gigabit-silicon-photonics-move.html>; August 21, 2016.

[4] The 300-pin multisource agreement, <http://300pinmsa.org>.

[5] The CFP multisource agreement, <http://www.cfp-msa.org>.

Optical Transmission Techniques for Layer 4 Networks

This appendix looks at the techniques used for optical transmission across Layer 4 networks—the telcos' metro and long-haul networks and data center interconnect links. The layering scheme is defined in Chapter 2, Layers and the Evolution of Communications Networks, while optical transport and data center interconnect are subjects discussed in Chapter 5, Metro and Long-Haul Network Growth Demands Exponential Progress.

For such transmissions, optical engineers seek to improve continually the amount of data that is sent over fiber. Engineers draw upon two techniques to increase the capacity of their optical transmissions: modulation and multiplexing.

Modulation is used to transmit data over a fiber by modulating particular characteristics of an optical "carrier" signal—a wavelength. The carrier's attributes that can be modulated, thereby encoding the data, include amplitude, phase, and frequency. The carrier is chosen as one best suited for transmission across the particular communication link. For dense wavelength-division multiplexing, the carrier signal is light—a lightwave—typically around 1550 nm that is suited for long-distance transmission over the fiber's C band (see Fig. A1.4).

The second technique used by optical engineers to boost capacity is *multiplexing*, which can be viewed as a form of combining. Wavelength-division multiplexing is the most well-known approach to increase the capacity of data sent over an optical fiber.

Optical multiplexing as described here is used to increase the carrier's data payload. One example exploits light's two polarizations to effectively send data along two independent paths in parallel, thereby doubling the carrier capacity. Another example of multiplexing is spatial multiplexing—combining light across space-which was introduced in Chapter 5, Metro and Long-Haul Network Growth Demands Exponential Progress.

In the book *Optical Fiber Telecommunications*, Verizon's Glenn Wellbrock and T.J. Xie describe three types of optical channel using modulation or modulation and multiplexing combined [1]. The optical channel is a construct to send a block of data over fiber.

A2.1 THE THREE CLASSES OF OPTICAL CHANNEL

The first optical channel, *Type 1*, is the simplest. Here the optical channel uses a single carrier or wavelength with no multiplexing. This was the approach used in the early days of optical transmission where the data rate increased continually using an optical modulator to either pass or block light, known as on-off keying, and direct detection was used at the receiver (see Section 5.5.1). The approach works fine for 2.5, 10 and 40-Gb data rates, but going to higher speeds is problematic as the modulator struggles to keep up (Fig. A2.1).

A *Type 2* optical channel tackles the modulator's limitation by adding multiplexing to the single carrier. For a single carrier, the three ways to perform optical multiplexing are polarization, space, and time. We focus on the first two only; time-based multiplexing has not been commercially exploited.

Light has two polarizations (see Fig. A2.2) that allow data to be sent along separate independent paths [2]. Using polarization, transmission capacity can be doubled or, alternatively, the same data can be sent at half the signaling rate, thereby relaxing the demands made of the modulator and the transmitter's and receiver's electrical circuitry but at the expense of more complex optical componentry. This is why Type 2's use of multiplexing improves on Type 1 by effectively halving the modulator's bandwidth.

Type 2 is already deployed commercially for long-distance transmission, as discussed in Chapter 5, Metro and Long-Haul Network Growth Demands Exponential Progress, and explained in the next section. Spatial multiplexing is another Type 2 scheme and is viewed as a long-term solution to overcome the nonlinear Shannon limit of fiber.

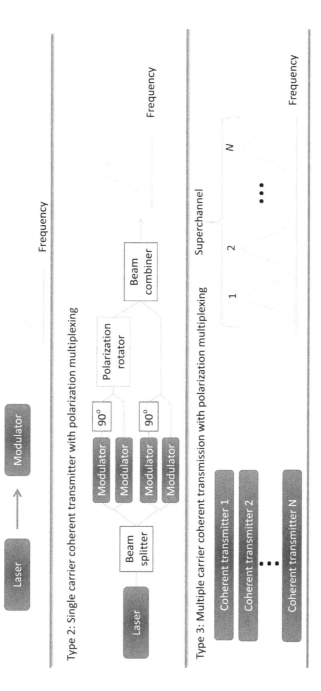

Figure A2.1 The three types of optical channel based on modulation and multiplexing.

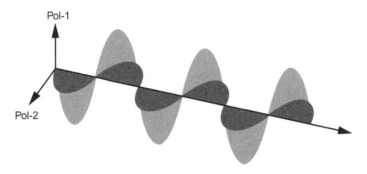

Figure A2.2 Two polarization states of light.

Table A2.1 Three Approaches to Building Optical Channels					
	Carriers	**Multiplexer**		**Modulation**	**Example(s)**
Type 1	Single	None: simple on-off keying		Variable	10 Gb
Type 2	Single	Variable		Variable	Polarization multiplexing for 100G (PM-QPSK) and Polarization multiplexing for 200G (PM-16QAM)
		Polarization	Time	Space		
Type 3	Multiple	Variable		Variable	Superchannels (carriers) for 400G
		Polarization	Time	Space		

The *Type 3* optical channel also uses modulation and multiplexing but exploits multiple carriers. There is only so much information a single carrier can hold—we talk about 400 Gb and even 600 Gb per wavelength in Section A2.5—so to expand capacity further, multiple carriers are used in parallel to form what is known as a superchannel.

The three schemes are shown in Fig. A2.1, while Table A2.1 summarizes their characteristics.

These optical channel categories are used to help explain how optical engineers are expanding optical transmission capacity.

A2.2 SINGLE-CARRIER 100-GB TRANSMISSION WITH COHERENT DETECTION

Toward the end of the last decade, there was much industry debate as to whether 10-Gb wavelengths should be succeeded by the next natural speed progression—40 Gb—or whether it would be prudent, albeit more challenging, to go straight to 100 Gb/s.

Until the debate was resolved in favor of 100 Gb [3], more complex modulation schemes than simple on-off keying were adopted to achieve 40-Gb data rates. The modulation schemes exploited the amplitude and phase of the light waves and included 40-Gb duo-binary, 40-Gb quadrature phase-shift keying (QPSK) and 40-Gb differential QPSK (DQPSK).

Despite going to 100 Gb/s transmission, the development of these more complex 40-Gb modulation schemes was not wasted work. It helped the industry alight on a Type 2 optical channel based on polarization-multiplexing, quadrature phase-shift keying (PM-QPSK), coupled with coherent detection for 100-Gb. Here two independent polarizations of light halve the signaling rate, while using QPSK, the phase of each of the two polarized streams is modulated, with each of the two components carrying data. Simply put, PM-QPSK encodes 4 bits per signal or symbol; the electronics is at 25 GBd/s as is the symbol rate but the data transported is at 100 Gb/s.

Choosing PM-QPSK allowed components already developed for 40-Gb optics to easily be used for the more demanding 100 Gb. This however results in more complex optical componentry. There are more paths—two polarizations, each with real and imaginary components—that results in a more complex and expensive transmitter (see Fig. A2.1 Type 2) and receiver design.

For long-distance transmission, forward-error correction codes are included alongside the data payload such that when 100-Gb transmission is mentioned, what is being referred to is the data payload only. The forward-error correction adds extra overhead bits so that in reality, the bit rate is between 128 and 140 Gb. This means that, in practice, the symbol or baud rate is between 32-35 GBd.

A2.3 IMPROVING SPECTRAL EFFICIENCY

Another issue optical designers fret about is the spectral efficiency of an optical channel.

The 40-Gb schemes of a decade ago occupied 100- and 50-GHz-wide dense wavelength-division channels; a spectral efficiency of 0.4 and 0.8 bits/s/Hz, respectively. For 100 Gb in a 50-GHz channel, the spectral efficiency improves to 2 bits/s/Hz. That is what PM-QPSK achieves: a higher spectral efficiency, which means more bits squeezed onto the fiber. But this must not be at the expense of overall transmission distance.

This is where coherent detection comes in. Coherent detection allows for the recovery at the receiver of all the information associated with the transmitted lightwave such as its phase and amplitude. This information allows digital signal processing techniques to be used to counter transmission impairments over the fiber. Coherent detection also uses a copy of the signal at the receiver to recover the data. Because a local copy used is "clean" rather than using a signal that has also traversed the fiber, coherent detection delivers an optical gain improvement and therefore improved transmission distances.

The latest coherent systems deliver 25 × the capacity-reach product versus traditional 10-Gb wavelengths that use Type 1 optical channels and direct detection, a notable advance that has taken a decade of engineering only to achieve [4].

A2.4 HIGHER-ORDER MODULATION

Designers have not stopped at 100-Gb PM-QPSK. Using more advanced modulation schemes, 6 bits per symbol is achieved using an 8-point constellation quadrature amplitude modulation scheme (8-QAM) and 8 bits per symbol using 16-QAM. These equate to 150- and 200-Gb data rates sent on a carrier. But sending more bits per second means the receiver has less scope for their correct recovery in the presence of noise, reducing the overall transmission distance that can be achieved. But the benefit is in improved spectral efficiency: doing the math, 150 and 200 Gb over a 50-GHz channel boosts spectral efficiency to 3 and 4 bits/s/Hz.

Optical designers also use pulse-shaping techniques at the transmitter. Shaping the sent signals helps squeeze adjacent optical carriers into narrower channels, better using the spectrum. For example, a 100- or 200-Gb signal can fit in a 37.5-GHz channel rather than a 50-GHz one. Indeed, channel sizes are being segmented into more granular increments, such as 12.5 GHz and even 3.125 GHz, known as flexible grid. Designers have also explored a gridless approach where channels can be arbitrarily sized, not just integer multiples of smaller increments such as 12.5 or 3.125 GHz.

Telecom operator BT and equipment maker Huawei trialed in 2015 a Type 3 optical channel: a 15-carrier superchannel where each carrier held 200 Gb of data in a subchannel slightly wider than 33 GHz. The resulting spectral efficiency achieved was just under 6 bits/s/Hz [5]. Such schemes require changes to the rest of the dense wavelength-division multiplexing line system. For example, the reconfigurable optical add-drop multiplexers used to switch optical carriers between fibers at network nodes also need to handle such flexible channel spacing.

Current work is focused on making transponders—advanced transceivers for long-haul-adaptable by supporting several modulation schemes to maximize the transmitted data for a given transmission distance. Already coherent systems support polarization-multiplexing, binary phase-shift keying (PM-BPSK), PM-3QAM, PM-QPSK, PM-8QAM, and PM-16QAM. And new proprietary modulation schemes are also being introduced [6].

PM-QPSK is the de facto 100-Gb standard that supports transmission distances of several thousand kilometers without optical signal regeneration. The simpler PM-BPSK delivers 50 Gb per carrier while enabling even greater transmission distances. Such a scheme is typically used for the most demanding long-distance fiber routes such as pan-Pacific submarine links. Using PM-8QAM achieves 150 Gb per carrier, 1.5 × more than QPSK, and 2000-km-plus distances, twice the 1000 km of PM-16QAM.

A2.5 THE LEVERS USED TO BOOST TRANSMISSION CAPACITY

Given that the bulk of data center interconnect links are in a metro network, there is an advantage in maximizing the amount of data that can be carried by a carrier as long as it meets the reach requirements. Accordingly, QAM schemes using more than a 16-point constellation (16-QAM) are being explored such as 32-QAM and 64-QAM. But as mentioned, reach drops dramatically the higher order the modulation scheme used.

This approach can be seen as maximizing the bits per symbol—traveling along the y-axis, as shown in Fig. A2.3. The goal is to increase the number of bits carried based on a given symbol rate, for example 32 GBd/s which is used today at 100 Gb with PM-QPSK.

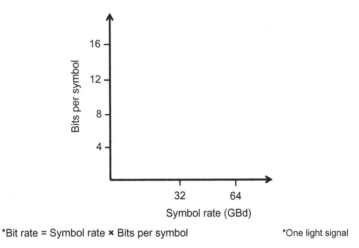

*Bit rate = Symbol rate × Bits per symbol   *One light signal

Figure A2.3 Two levers used to boost capacity: bits per symbol and symbol rate.

As mentioned above, PM-QPSK uses 4 bits per symbol. But the symbol rate is not limited to 32 GBd/s. Increasing the baud or symbol rate—moving along the x-axis—is an additional lever to increase the overall data sent by the carrier.

Optical transport systems have already been announced that use a higher symbol rate than 32 GBd. Cisco's NCS 1002 data center interconnect platform, discussed in Chapter 5, Metro and Long-Haul Network Growth Demands Exponential Progress, supports 250 Gb on a single wavelength using PM-16QAM and 40 GBd. Ciena's WaveLogic Ai coherent DSP-ASIC supports two baud rates: 35GBd/s and 56GBd/s [7]. And work has started among component makers to double the symbol rate to 64 GBd/s. That would mean each of the two polarization's real and imaginary components of PM-QPSK would carry on the order of 64 Gb, resulting in a total data rate of 200 Gb on a carrier. Using PM-16QAM, the payload would achieve 400 Gb per wavelength, and 64-QAM would result in a 600-Gb wavelength [8].

Once systems are able to support 64 GBd, 400-Gb transmission using 16-QAM will revert from a Type 3 optical channel to a Type 2 one. That is because instead of using two carriers, each carrying 200 Gb (and two transmitters and two receivers), only one channel will be needed. However, it should be noted that doubling the symbol rate doesn't improve spectral efficiency: going to higher speeds increases the carrier width in the frequency domain.

But before the baud rate can be doubled to 64 GBd, vast improvements in component performance are needed such as doubling the modulator's bandwidth and increasing the speed of the photodetectors and the associated modulator driver and receiver amplifier electronic components. The analog-to-digital converters that sample each of the four streams of the PM-QPSK received signal and turn them into digital bit streams for processing also need to operate at twice their current rate.

To recover a signal, it needs to be sampled at twice its highest-frequency component. At 64 GBd this equates to an analog-to-digital converter working at up to 128 giga-samples per second. In other words the converter must take a sample—representing the signal digitally—once every eight-billionth of a second. Equally the coherent digital signal processor must implement its various algorithms on a doubled-rate data stream.

Optical designers thus have at their disposal two knobs—modulation (bits per symbol) and symbol rate—to maximize data rate for the required distance. And this is what the systems vendors are up to: working to develop line-side optical transmission designs that increase the symbol rate beyond 32 GBd toward the goal of 64 GBd, while adding further modulation scheme options between the extremes of BPSK and 64-QAM.

But there is a third knob, as we have seen, to increase capacity. Once a single carrier is optimized, multiple copies can be used in parallel to create a Type 3 optical superchannel. This can be seen as adding an extra dimension to the bit and baud rates. This z-axis is shown in Fig. A2.4.

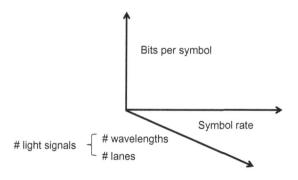

Bits per symbol

Symbol rate

light signals ⎰ # wavelengths
              ⎱ # lanes

Bit rate = Symbol rate × Bits per symbol × # light signals

Figure A2.4 Three levers used to boost capacity: bits per symbols, symbol rate, and # light signals.

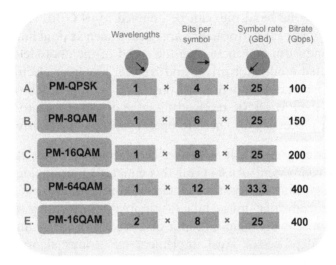

Figure A2.5 The knobs available to effect different data rates, distances, and spectral efficiencies.

The Type 2 single-carrier multiplexed approaches highlighted in Fig. A2.5 are Examples A, B, C, and D.

Examples A, B, and C use increasingly complex modulation schemes to increase the data contained in the channel, but this is at the expense of distance. In Example D, PM-64QAM is used. Here the equivalent of 6 bits per symbol are carried on each polarization. If 25-GBd signaling is used (assuming no overhead bits), the result would be a 300-Gb bit-rate using one carrier.

To achieve 400 Gb, a slightly faster symbol rate—33.3 GBd—is needed. This is more demanding than 25 GBd but nowhere near as demanding as the doubling needed to achieve 400 Gb on a single carrier with 50 GBd and PM-16QAM. But again, adopting 64-QAM compromises transmission reach further.

Example E adopts a Type 3 optical channel by bonding two carriers together—each carrying 200 Gb—to form a simple two-carrier 400-Gb superchannel.

REFERENCES

[1] Xia TJ, Wellbrock G. Commercial 100 Gbit/s coherent transmission systems. Optical fiber telecommunications VIB: systems and networks. 6th ed. Oxford, UK: Academic Press; 2013 [chapter 2].

[2] Kasap SO. Optoelectronics and photonics: principles and practices. Boston, MA: Pearson; 2013.

[3] Jagdeep Singh's Infinera effect. Gazettabyte, <http://www.gazettabyte.com/home/2009/12/27/jagdeep-singhs-infinera-effect.html>; December 27, 2009 and High fives: 5 Terabit OTN switching and 500 Gig super-channels. Gazettabyte, <http://www.gazettabyte.com/home/2011/9/15/high-fives-5-terabit-otn-switching-and-500-gig-super-channel.html>; September 15, 2011.

[4] Next-generation coherent adds sub-carriers to capabilities. Gazettabyte, <http://www.gazettabyte.com/home/2016/1/24/next-generation-coherent-adds-sub-carriers-to-capabilities.html>; January 24, 2016.

[5] BT makes plans for continued traffic growth in its core. Gazettabyte, <http://www.gazettabyte.com/home/2016/1/19/bt-makes-plans-for-continued-traffic-growth-in-its-core.html>; January 19, 2016.

[6] Heading off the capacity crunch. Gazettabyte, <http://www.gazettabyte.com/home/2015/4/2/heading-off-the-capacity-crunch.html>; April 2, 2015.

[7] Ciena unveils WaveLogic Ai, Ciena Insights, <http://www.ciena.com/insights/articles/Ciena-unveils-WaveLogic-Ai.html>; October 31, 2016.

[8] Finisar introduces 64 Gbaud high bandwidth integrated coherent receiver, press release, <http://files.shareholder.com/downloads/FNSR/2964083257x0x908619/1739ca15-941a-458f-877d-191cfc13ef6d/FNSR_News_2016_9_19_General.pdf>; September 19, 2016.

INDEX

Printed in the United States
By Bookmasters